Book of Texas Moths

GIDEON LINCECUM NATURE AND ENVIRONMENT SERIES

SPONSORED BY JERRY B. LINCECUM AND PEGGY A. REDSHAW

Book of Texas Moths

GARY CLARK
and KATHY ADAMS CLARK

Photographs by JOHN TVETEN

TEXAS A&M UNIVERSITY PRESS
College Station

First edition

∞ This paper meets the requirements of ANSI/NISO Z39.48–1992
    (Permanence of Paper).
Binding materials have been chosen for durability.

Manufactured in China through Martin Book Management

Library of Congress Cataloging-in-Publication Data
Names: Clark, Gary, 1943– author. | Clark, Kathy Adams, 1955– author. |
    Tveten, John L., photographer.
Title: Book of Texas moths / Gary Clark and Kathy Adams Clark; photography
    by John Tveten.
Other titles: Gideon Lincecum nature and environment series.
Description: First edition. | College Station: Texas A&M University Press, [2025] |
    Series: Gideon lincecum nature and environment series | Includes index.
Identifiers: LCCN 2024054296 (print) | LCCN 2024054297 (ebook) |
    ISBN 9781648432361 | ISBN 9781648432378 (ebook)
Subjects: LCSH: Tveten, John L. | Moths—Texas—Identification. | Moths—
    Texas—Pictorial works. | Naturalists—Texas—Biography. | Moths—Research. |
    Nature observation. | BISAC: NATURE / Animals / Butterflies & Moths |
    NATURE / Reference | LCGFT: Handbooks and manuals. | Biographies.
Classification: LCC QL551.T4 C53 2025 (print) | LCC QL551.T4 (ebook) |
    DDC 595.7809764—dc23/eng/20241207
LC record available at https://lccn.loc.gov/2024054296
LC ebook record available at https://lccn.loc.gov/2024054297

To Shannon Davies, former director of Texas A&M University Press,
who sparked the idea for this book

Contents

Book of
Texas Moths

Introduction

Adult Moonseed Moth roosting on a sunflower.

The Life of John Tveten

This book is based on the original fieldwork of John Tveten, PhD (1934–2009), whose life stands as a testament to what a person can discover by careful and systematic observation of the natural world. Born with a love of nature and educated as a research chemist, John originally worked for a major oil company but would later leave the company to fulfill his dream of becoming a full-time naturalist and nature writer. Beginning in 1978 until his untimely death in 2009, John wrote meticulous field notes about nearly every bird, butterfly, mammal, reptile, dragonfly, and moth he encountered. Neither did any blade of grass, any weed, any bush, any flowering plant, or any tree get past his inquiring eye or field notes. John observed nature with the joyful curiosity of a child and with the strict discipline of a scientist.

John began writing field notes on loose-leaf paper but later used a form designed by his wife Gloria, who shared his enthusiasm for documenting observations of the natural world. The form included the date, location, common name, and scientific name of each critter he had observed. An example of one of his forms for moths is shown here:

_L 2456_   Common Name: _(BLANCHARDS APOTOLYPE)_

Working Name: Scientific Name: _APOTOLYPE BLANCHARDI - FRANC._
TOLYPE MOTHS

_____   Family: _LASIOCAMPIDAE/MACROMPHALIINAE_

Location: Habitat/Foodplant:
TX   CAMERON CO.   _CP WALL NEAR COCOON_
LOS FRESNOS
(BEBB WILLOW)
Date: _10 MAY 02_   _(SALIX BEBBIANA)_

Reference: _____

Photos: _____

Specimens: _____

Date	Description	Photo
5-10	_Pair of freshly ... Lasiocampid moths mating ... Female (?) cocoon nearby._	
5-31 (Menu)	_Female began laying eggs ... Eggs hatch - very tiny ... large larvae. ... Aspen - no / Black - no / ... - no / Plantain - no / Dandelion - ... / Petals - at once / ... at once / [... preferred]_	
7-5-02	_...on a broad-leaved willow / ... larvae began pupating ... two ... large ... 5-7 mm long_	
7-10-02	_Largest larvae open on side of jar - probably ♀'s_	

John constantly collected both butterfly and moth caterpillars. Despite John's encyclopedic knowledge of butterfly caterpillars, he was not equally knowledgeable about moth caterpillars. Thus began his quest to study moths. When John discovered an unknown moth caterpillar, he would collect it along with the host plant (the caterpillar's food plant), put both in a jar, and wait to examine and classify the species of moth that eventually emerged. He followed that routine of discovery even with identifiable moth caterpillars. Jars or other containers containing both known and unknown caterpillars along with their host plants lined shelves, bookcases, windowsills, and kitchen counters in the Tveten home. John then documented the development of moth caterpillars as they feasted on a plant, grew larger, formed cocoons, and eventually emerged as winged moths, many of which were as beautiful as butterflies. That led John to create a presentation titled *Butterflies of the Night* to educate audiences about the beauty of often overlooked moths.

John carefully linked his moth photographs to corresponding field notes detailing the when, where, and what of moth observations. He planned to use his notes to write a book about Texas moths and had presented an outline to the editors at Texas A&M University Press. Death suddenly silenced his pen.

My wife, Kathy Adams Clark, and I wanted to complete John's dream of a moth book. He and his wife Gloria were our friends and mentors. We heard stories about John's moth discoveries and examined many of the specimens he had captured and documented. We also enjoyed telephone conversations with John about moths among his collection that were emerging, hatching, or metamorphosing. We know John's voice. We know his mind.

We are aware that completing John's book would necessarily leave out newly discovered moth species and new information about the distribution of moths. But our goal was to remain faithful to John's original research while including essential updates about species he'd observed. Even though we have written about John's careful documentation on the distribution, life history, and identification of almost 100 Texas moth species, we have not intended the book to be a field guide to moths. That wasn't possible because knowledge about moths has expanded since John's passing. Instead, we wrote a book that takes readers into the mind of a disciplined naturalist as he conducts original research to document the kinds of moths in Texas. Getting into the mind of a consummate naturalist on a quest to learn about moths is a transformative experience for anyone who enjoys nature.

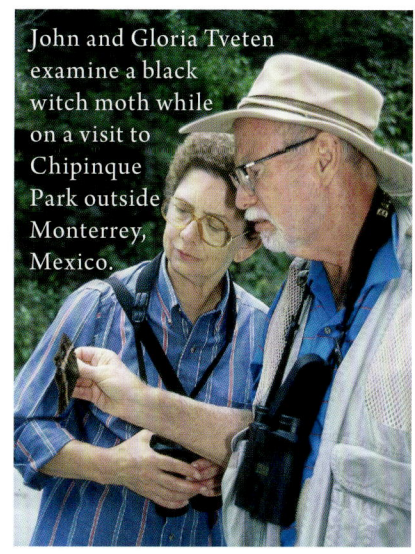

John and Gloria Tveten examine a black witch moth while on a visit to Chipinque Park outside Monterrey, Mexico.

Ailanthus Webworm Moth
photographed by John Tveten.

Looking Back on John Tveten

After John Tveten's passing in 2009, those who knew him both personally and professionally were devastated. I wrote the following words, which appeared in *Texas Parks & Wildlife* magazine, to describe his life and work.

THE CONSUMMATE NATURALIST

I never knew a naturalist quite like John Tveten, who passed away on October 12, 2009, just before his 75th birthday, after a brief bout with cancer. John was brilliant but humble, knowledgeable but always studious, dedicated to careful observations but always ready to share them. Walking with him outdoors was like walking with a talking volume of field guidebooks for plants and critters. Shannon Davies, John's editor at Texas A&M University Press, knows such a walk.

In telling the story of walking with John in a vacant lot at Rockport, Davies recalls, "When I walked with John, I learned there was no such thing as a 'vacant' lot. In that tromped-down, overmowed, sorry patch of earth, John saw tiny butterflies on tiny wildflowers growing in the grass with the ants, flies, bees, bugs and minuscule snails. Later, I learned that not only did John know everything, he could write books about everything."

Yes, John knew everything. He was a consummate naturalist in the tradition of such legendary 20th-century Texas naturalists as Henry Attwater (1854–1931), Roy Bedichek (1873–1959), and Harry Oberholser (1870–1963). As did his naturalist predecessors, John wrote detailed accounts of nature, including books and articles on birds, butterflies, wildflowers, and coastal ecology. However, unlike most naturalists before him, John was an accomplished nature photographer whose pictures illustrated not only his own publications but also scores of books and magazine articles. His photographs, for instance, provided compelling visual documentation in David Schmidly's landmark book *The Mammals of Texas* (University of Texas Press, 2004). His photographs also mesmerized all of us who attended his deeply informative lectures at nature festivals, nature clubs, schools, and museums.

Often in the background but hardly out of the way was John's wife, Gloria, who was also an accomplished naturalist. She assisted John in all of his writings long before her name appeared with his name in publications. John never failed to acknowledge her and her help. She always sat at the back of the room to operate the slide projector during John's stirring lectures, and her accompaniment on the projector freed John to soar as a lecturer with his enticing rhetoric.

One of his lectures at the Houston Museum of Natural Science in the late 1970s profoundly influenced me.

"You don't have to go to faraway places to see the wonders of nature," John intoned as his eyes glistened with excitement. "The wonders of nature are right here in Houston." Gloria clicked the slide projector. On the screen, in glorious splendor, was a huge flock of snow geese arrayed across a blue sky, set off against the backdrop of an oil refinery. John's resonant voice boomed. "I've seen beautiful sunsets all over the world, but can you beat this sunset?" Gloria clicked the slide projector again. And on the screen, the richly hued orange disk of the sun, with gulls silhouetted against it, touched the sea at twilight off Galveston Island.

John dazzled the rest of the audience and me that evening with nearly a hundred slides of colorful birds, resplendent butterflies, and gorgeous wildflowers all photographed in the greater Houston area. His face was beaming when he said, "The Houston area is a wonderful place to enjoy the wonders of nature."

In that lecture, John reminded me, a native-born Houstonian, that some of the world's richest natural treasures lay right at home. From that point on, I began working hard with local nature clubs and conservation organizations to help build interest in the wildlife of Houston and in wildlife throughout Texas. John gave me the inspiration and fueled my enthusiasm to do that work. Fortunately, his and Gloria's book *Nature at Your Doorstep: A Nature Trails Book* (Texas A&M University Press, 2008) documented the joy of nature where we live, and we can hope it will fire the enthusiasm even in people yet unborn to conserve natural treasures in their hometowns.

John's original hometown was Morris, Minnesota where he was born on October 16, 1934. He began life as a naturalist at an early age by building a large collection of butterflies in and around his home state. While still a boy, he corresponded with leading lepidopterists around the world and became so well known that people from Europe and Japan called upon him after World War II to help resupply butterfly specimens to war-ravaged natural history museums.

In college, John studied chemistry and earned a PhD in organic chemistry from the University of Illinois in 1960. Gloria, whom he had married in 1958, earned a master's degree in mathematics that year from the same university. After graduation, the couple traveled to Baytown, where John took a job as a research chemist at ExxonMobil (then known as Humble Oil and Refining Co.). Gloria joined the faculty of Lee College in Baytown as a professor of mathematics.

John had been pursuing a hobby as a naturalist and nature photographer during his tenure at the refinery. One day, while sitting in an office meeting, he found himself more interested in a spider creeping along the conference table than he was in the momentous chemical discussions at hand, so he resigned from the company in 1973 to become a full-time nature photographer and writer. He also became a nature tour leader for such organizations as the Smithsonian Institution's travel program, the National Audubon Society, the Houston Museum of Natural Science, and the Spring Branch Nature Center (now called the Robert A. Vines Environmental Science Center) in Houston.

His photographs began appearing in hundreds of magazines, books, calendars, and filmstrips. His articles began showing up in state and national magazines, including *Texas Parks & Wildlife* magazine and *Smithsonian* magazine. As the years progressed, John produced numerous books, among them *The Birds of Texas* (Shearer Publishing, 1993) and, along with Gloria, *Wildflowers of Houston and Southeast Texas* (University of Texas Press, 1997) and *Butterflies of Houston and Southeast Texas* (University of Texas Press, 1996).

When John died, he was working on a definitive book about moths, a project for Texas A&M University Press. In typical fashion, John's research for the book included raising caterpillars to learn firsthand the life cycle and identification of moths.

Many people first became familiar with John through his newspaper column, "Nature Trails," that began in the *Houston Chronicle* in 1975 and ran until March 1999. The column was first written under John's byline, but later under a joint byline with Gloria. I devoured that column with the eagerness of a hummingbird devouring nectar. And like nectar to a hummingbird, John's columns energized me as a naturalist.

Legions of nature buffs, including birders, butterfly watchers, and wildflower enthusiasts, can trace their inspiration and early teachings to John. He encouraged in people not only a knowledge of nature but also an appreciation.

For example, when my wife, Kathy Adams Clark, began her career as a professional nature photographer, she turned to John to guide her with his legendary photographic skill. John taught her to know a critter or a flower and to know it well before taking a picture. Kathy now drums that lesson into other photographers.

Tributes to John have been pouring in since his death. Kathy, who was at his bedside when he passed, said, "All of us who called him a friend will remember his strong love of this planet and his optimistic spirit. He was a naturalist first and a photographer second. He always said the critter was more interesting than the camera."

John's former neighbor and Texas naturalist David Dauphin said, "John's books, field guides, newspaper articles, field trips, and programs filled us with knowledge, the desire to see more and the need to savor nature slowly. John was a good friend, a kind man, a gentle man, a loving husband and father. I don't ever remember a frown on his face." Tom Collins, once the co-compiler for the Freeport Christmas Bird Count, said, "I remember that John didn't just give the typical field guide discussion of birds. His words brought them to life and made you want to know more about them."

Greg Lasley, author of *Greg Lasley's Texas Wildlife Portraits* (Texas A&M University Press, 2008), said, "I'm proud to have called John a good friend for more than 30 years. He was one of the finest persons it has been my privilege to know, and I will miss him greatly. The writings about birds, butterflies, moths and other natural history subjects that John and Gloria produced over the years have enriched us all and leave a legacy for us to cherish."

Kenn Kaufman, internationally known author of bird and butterfly guides, said, "I've never met a finer naturalist than John. His knowledge of everything in the outdoors, and his enthusiasm for it, were just extraordinary, but despite that he was amazingly humble."

Texas naturalist Tony Gallucci composed a poem, which in part reads:

John spent his lifetime
First and foremost as a teacher
Sure he took photographs
But they were framed as visual lessons
Sure he raised caterpillars
But not for himself
Sure he wrote books
But to spread what he had learned himself

As for me, I thought of John as a scientist in mind and a poet in heart. He scrutinized nature with the inquisitive but exacting mind of a scientist. For example, in *Butterflies of Houston and Southeast Texas*, John wrote with the precision of a scientist about the complex family of longwing butterflies: "The concepts of family, genus, and species, after all, are artificial human constructions devised for our convenience. They help us indicate relationships among populations. The various butterfly populations, however, do not adhere to the strict order we impose."

But he could also engage us with the heart of a poet as when he described his and Gloria's lifetime of observing birds in *Our Life With Birds: A Nature Trails Book* (Texas A&M University Press, 2004): "We enjoy seeing uncommon birds, but we also enjoy seeing common birds doing uncommon things. And, most of all, we simply enjoy birds being birds uncommonly well."

I believe that John will be ranked among the greatest naturalists. His knowledge was encyclopedic, and his generosity and vitality in sharing that knowledge were without equal. Over the years, whenever I called on him for help or advice, he was always generous, always helpful, and always excited to talk about natural wonders. John will live in my memory as a great man for his knowledge and an even greater man for his heart.

About This Book

The *Book of Texas Moths* draws entirely from John Tveten's original research. His extensive field notes have been transcribed by Kathy Adams Clark and will be stored with the original field notes at Texas A&M University in College Station, Texas.

As much as possible, Kathy and I have retained John's voice by weaving sentences and paragraphs from his field notes and have incorporated his descriptive phrases into as many species accounts as possible. To be faithful to John's work, we have updated common and scientific names of moth species where necessary while retaining the historical names. We have included the distribution, or range, in Texas based on 2020 sources. For historical purposes, we have included the locations where John collected moth specimens.

Our thanks to Janice Braud, Patti Edens, Sharron Jay, Stuart Marcus, and Farrar Stockton. All contributed to help bring Tveten's work to life.

More than anything, this book comes from the mind of the greatest naturalist we have ever known.

Snowberry Clearwing Moth. Moths come in many shapes and sizes.

Plume Moth has an unusual shape for a moth.

All about Moths

How often do we pass by moths under a porch light or fluttering under a streetlamp without giving them a second glance? Yet moths, with their intricate patterns and curious coloration, are equal in beauty to butterflies. What's more, they far outnumber butterflies.

Nearly 11,000 species of moths are found in the continental United States, compared with about 765 species of butterflies. Moths are so plentiful that many have yet to be given common English names. The ones that do have English monikers certainly have enchanting ones, like the luna moth (*Actias luna*). This elegant, softly hued green moth has a 3- to 4-inch wingspan with two long streaming tails. In early spring, it has deep purple lining on the edge of its forewings.

Moths are in the same order as butterflies, Lepidoptera. But moths have threadlike or feathery antennae whereas butterflies have clubbed antennae. Most moths except for a few diurnal moths fly by night, whereas butterflies fly by day.

Ecologically, moths pollinate multiple plant species. They are also an important link in the food chain because their larvae or caterpillars provide nourishment for birds, and their flying form is chow for bats and screech owls.

Moths, like this Polyphemus Moth, have feathery antennae.

Double-Lined Prominent moths have threadlike antennae.

Common grackle eating a Pawpaw Sphinx larva.

Spider feeding on a caterpillar.

Corn Earworm Moths can be pests but bats can keep them under control.

Some moths are surely pests. For example, corn earworm moths can cause major crop damage. However, if we encourage habitat for bats, they will keep the pesky moths under control.

Raising Moths in Captivity

John Tveten was a trained scientist and believed in primary research. As a naturalist, he took pride in his observations in nature. That's why he obsessively recorded his observations in notebooks. Those observations are the basis for this book.

Most of John's observations come from captive species. He would find a butterfly or moth larva in the field, put it in a jar, and take it home. He might put an adult moth in a jar or bag to see if it would lay eggs. Then he'd release the adult and raise the eggs to maturity.

John's scientific discipline prepared him for watching the metamorphosis of these specimens with his own eyes, taking notes, and jotting down observations.

Larva stage of a Hieroglyphic Moth that Tveten raised to the moth stage.

We both visited the Tveten home over the years and saw the shelves of jars containing specimens. His favorite jars once held peanut butter but were repurposed as insect cages. He often couldn't take a phone call because something interesting was hatching or emerging.

In physics, there's the observer effect. The observed system is disturbed by the act of observation. We don't think that the butterflies and moths John observed were disturbed by his observations in captivity.

Adult Hieroglyphic Moth.

A specimen might have eaten something different than in the wild but overall, the value in Tveten's work comes from his observations.

We've raised moth larva to maturity under the guidance of John and other naturalists. This is fun, educational, and can be a huge responsibility. If you choose to raise butterflies or moths, please do so responsibly.

John Tveten photographed Indian Meal Moth larvae he found in a bag of bird seed.

About the Photos in This Book

John Tveten, PhD was known as one of the foremost nature photographers in the United States during his career. From 1965 onward photo editors could call John and find the photo they needed for their magazine, newspaper, or film strip.

All the photos in this book were taken by John. A behavior, color morph, instar stage, egg, or adult moth is illustrated with a photo taken by John when he wrote his field notes. Every slide in his extensive collection was numbered to correspond to his field notes. That numbering system had helped us illustrate the text with the moth John raised or encountered in the wild.

Our thanks to Texas A&M University Press for keeping John's slides in a secure vault since his death. This book is richer because we can match the slide to the species descriptions.

What Are "Hodges Numbers"?

Ronald Hodges created a number system in 1983 while editing *The Check List of the Lepidoptera of America North of Mexico, including Greenland.* He assigned a separate number to distinguish each butterfly and moth on the checklist, namely because many moths and tropical butterflies are

Adult Black Witch moth.

designated solely by their scientific names and have no common names. For example, a tiny moth with the scientific name *Idia julia* has no common name and is instead designated with Hodges #8328. Other moths in the *Idia* genus with no common name are also designated with Hodges numbers.

The scientific classification of certain moths may change based on new evidence, which results in some moths in a taxon (taxonomic group) being "lumped" together as the same species. When a moth in a taxon is "split" into separate species, its Hodges number gets a decimal point. For example, Schaus' Tussock moth having been split from Davis' Tussock moth is now Hodges 8205.1.

Another numbering system besides Hodges is the P3 system, named for Greg Pohl, Bob Patterson, and Jonathan Pelham.

John used Hodges numbers in his field notes. We used those numbers because they're commonly referenced by lepidopterists and amateur moth enthusiasts. Hodges numbers also may be referred to as Moths of North America (MONA) numbers.

Hodges numbers are regularly used by the moth photographers group at the Mississippi Entomological Museum at Mississippi State University and by the iNaturalist listing site, both of which provided us with valuable references.

Moths

Adult Evergreen Bagworm moth.

Evergreen Bagworm *Thyridopteryx ephemeraeformis* 0457

Male adults have a robust black body with clear wings.

Bagworm larvae live in a small, hanging case covered in tiny pieces of wood or grass. Females live in their larval case.

In 1978, John Tveten wrote, "During the fall I found huge numbers of bagworm cocoons on many different trees. None had active larvae, and none seemed viable (of several that were opened).

"Some of the cocoons were photographed, but nothing more was done. I will work on these more next year and start earlier.

"The ornamentation varies greatly on the cocoons. I do not know how many species are involved."

Tveten gathered several cases from a bottlebrush bush and observed the bagworms "growing and expanding their cocoons." In another instance, "numerous cases found on gate. Many attached; others still mobile."

An adult emerged 35 days after it was collected. Tveten noted that the black male had pectinate antennae with long branches. It had wide, clear forewings.

Range in Texas: Eastern half of the state and into the Panhandle

When Found in Texas: All year

Food Plant: Bald cypress, locust, oak, bottlebrush, willow, lichen

Bagworm cocoons can be covered in different types of vegetation.

Bagworms live in the cocoon.

Ailanthus Webworm Moth *Atteva aurea* **2401**

The adult of this moth looks like a beetle. Its sleek, long wings are held tight over the body. Rusty-orange forewings are banded with white edged in thick black. Long black antennae and black legs held away from the body add to the beetle-like appearance.

Tveten found adults on three occasions but never noted the eggs or larvae.

Larvae are small, dark webworms.

Adults are ½ inch long. Specimens were found with blacklights in the Rio Grande Valley and Texas Hill Country.

Range in Texas: Throughout the state except for the El Paso area

When Found in Texas: All year

Food Plant: *Ailanthus altissima*, also known as tree-of-heaven. This is an invasive plant that crowds out other natives.

Adult Ailanthus Webworm Moths resemble a beetle.

Adult Oblique-Banded Leafroller.

Oblique-Banded Leafroller *Choristoneura rosaceana* **3635**

The brown mottled forewing on this moth is only ½ inch long. Wings are held against the body forming a rounded shape that looks a bit like a pudgy fish. The apex of the forewing flares out forming the fish's tail. Long, narrow antennae are held back against the wings. Tveten wrote that this is a "beautiful tapestry-patterned moth."

True to their name, leafrollers roll sections of a leaf together. A leaf of a dogwood, for example, might be rolled lengthwise and stitched together with silk.

The small larvae are very pale green with green heads.

At 15–22 mm long, larvae were emerald green with black/brown head and collar. At this stage, Tveten noted that they were "heavy bodied."

At 23–30 mm long, larvae were jade green with an orange-brown head. Two pairs of hairs were on the side of each segment. Tveten noted that larvae appeared to be different species because they looked so different.

Tveten asked in his notes if this species could have dimorphic larvae due to observations of color and size variations.

Pupation occurred in the bottom of a jar in captivity. In another instance, a pupa formed inside a tightly rolled leaf. Another was in a loose web on the jar. One pupated in a web at the top of the jar and others fastened their rolled leaf to the glass.

Adults emerged four to eight days after forming a pupa. Adults found together and raised together were lighter and darker but with overall same pattern.

Range in Texas: East Texas along a line from Dallas to Austin and San Antonio

When Found in Texas: March to October

Food Plant: Hackberry, flowering dogwood, black cherry, basket oak, black walnut, white oak, curly dock, white-flowered smartweed

The head of an Oblique-Banded Leafroller larvae darkens with age.

Adult Southern Flannel Moths.

The hairs on the larvae of a Southern Flannel Moth can cause a nasty sting.

Southern Flannel Moth *Megalopyge opercularis* **4647**

A flannel moth could be described as "fuzzy." Adults have golden-cream colored wings, body, and legs. The thorax and forewing base are darker. Legs and body are covered in fuzz. Wings are soft and appear velvety.

Tveten documented this species often from his home in Baytown to the Rio Grande Valley and into the Texas Hill Country.

Larvae of the Southern Flannel Moth are known as asps or stinging asps. "They sting like fury," Tveten wrote. "I accidentally brushed one in removing another and my finger hurt and burned for 10 hours."

Eggs were collected from females placed in a jar or bag with food. One female laid eggs in short strings and covered with hairs from her body until almost invisible. The eggs were pale yellow ovals.

Larvae emerged 15 days after the eggs were laid. The tiny "slugs" were pale tan with long hairs. As each grew, the hairs become denser to form the asp profile. Hair colors were described as gray-brown, rich tan, rusty tan, light gray, and gray.

The larvae of Southern Flannel Moth can be a variety of colors.

Within 10 days the larvae were 5 mm long and in their third instar. "Many have 'tangles' of hair in little globs on dorsum" Tveten wrote.

Pupation took place "over next couple of weeks." Several spun their cocoons on forked twigs. Cocoons had a "thin veil over trapdoor end."

Emergence was six weeks to four months later. A specimen observed outside spun in early November and emerged mid-December. Another specimen laid eggs two days after emerging.

Tveten found a large larva feeding on Chinese tallow near Wallisville, Texas. "These are the first caterpillars I have ever seen on tallow; they were seen actually eating leaves."

He found another unusual item when seven parasitic flies emerged from one cocoon. He wrote, "I thought flies were usually single."

Range in Texas: East Texas to Dallas and the Hill Country

When Found in Texas: March to December

Food Plant: Black-brush acacia, willow oak, post oak, live oak, sweetgum, Chinese tallow, and canna. This species feeds on the surface of the leaf, skeletonizing it.

Pupa of a Southern Flannel Moth can be hard to see on a forked twig.

Copied from John Tveten's September 1, 1983 "Nature Trails" column:

Most people in Houston are familiar with the stinging creature we call the "asps," an inch-long, tear-drop shaped insect clothed in a flowing coat of silky hair. Few realize, however, that beneath that bur coat and the hidden venomous spines is nothing but a caterpillar, the larva of a harmless moth.

I put the name "asp" in quotes because historically it has also been applied to several venomous snakes—including the Egyptian cobra that supposedly killed Cleopatra some 2, 000 years ago. Our stinging caterpillar, properly called the puss-moth caterpillar, is by no means that formidable, but it does provoke a strong reaction in some of its victims.

My first encounter with the puss caterpillar occurred many years ago when one fell down the back of my neck as I was pruning trees. Not realizing what was crawling inside my shirt, I squashed it with my hand and was immediately rewarded with a searing, burning pain. For several hours, I didn't feel well.

First aid often recommended for an "asp" sting consists of applying ice to the area followed by a paste of baking soda to neutralize the venom. Just as with many other insect stings, some people are more seriously affected than others. Occasionally one may require a doctor's care.

The inch-long puss-moth larva is scientifically known as *Megalopyge opercularis*, a name that in print is longer than the insect itself. It is not the only stinging caterpillar, for there are several others found quite commonly in the area. The large, spiny, green Io moth larva causes an itching rash much like that of stinging nettles, but the irritation usually disappears in a few minutes. The Spiny Oak-Slug, the Saddleback caterpillar, and the Hag moth larva all possess this effective defense. Megalopyge, however, seems to be the worst of the lot. By far the greatest majority of caterpillars, of course, are totally harmless and can be handled with impunity; there are only a few that are best left alone.

Puss caterpillars feed on the leaves of several different trees but are particularly fond of the many species of oaks. Cloaked in hair of brown, gray, or light tan, they are well camouflaged among the fallen leaves of autumn, the season in which they appear so commonly.

Southern Flannel "asp" or puss caterpillar.

When ready to pupate, the "asp" spins a hard little cocoon with a ridge across the center. A beautifully constructed, hinged trapdoor at one end can be pushed open from inside to allow the emerging moth to escape. It is this operculum, or trap door, that gives the insect its scientific name *M. opercularis.* Cocoons can be found by the score on tree trunks and branches or on fence boards and under the eaves of houses. The caterpillars also seem to favor the rough mortar between the bricks of buildings to build their tough silken hibernacula.

The adult moth that will eventually emerge from the pupal shell in spring and push open the operculum is small and yellow-buff in color. It, too, is clothed in long hairlike scales, but there are no venomous spines concealed within those hairs. The moth is completely hairless. Its sole purpose in life is to lay the eggs that will produce another generation of unpleasant tear-drop "asps" that crawl about in luxurious floor-length fur coats.

Adult White Flannel Moth.

White Flannel Moth *Norape virgo* **4650**

This large white moth is "fuzzy" but only on the thorax. White wings are velvety with a white fringe on the hindwing. Legs are black and antennae are pink.

In October, years apart, Tveten found larvae by the dozens on blackbrush acacia. He described them as 1½ inch long, fat, and stringlike. They had a harlequin pattern with an ivory background, clusters of yellow spines, and blue circles around each cluster. "Most have yellow spines: a few have orange or pinkish spines."

Early larvae stage of White Flannel Moth.

Tveten wrote, "There are some on most of the trees. In spite of this gaudy pattern, they are hard to see as they lie on the twig."

Cocoons were small and papery. Another sample of cocoons were described as "spun hard, tight little cocoons."

Adults emerged in May and June seven months after pupating.

Range in Texas: Corpus Christi to the Rio Grande Valley per Tveten. Current records Central Texas to Central Texas Coast.

When Found in Texas: April to October

Food Plant: Black-brush acacia

Larvae are gaudy and change color as they age.

Spiny Oak-Slug Moth   *Euclea delphinii*   **4697**

This is a small moth measuring ⅓ to ½ inch long. Wings are brown with a varying amount of lime-green markings. Body and legs are fuzzy, per Tveten.

Tveten found these slug caterpillars from September to November. One was orange and green, but another was all green.

Pupation was approximately a month after placed in captivity. The pupa was a small black oval.

Adults emerged three to five months later. Those found in the fall emerged in early spring. One specimen had green wings with brown borders. Another had "wings mostly green with brown spots."

Range in Texas: East Texas from Dallas/Fort Worth and Austin east

When Found in Texas: March to November

Food Plant: Hop hornbeam, willow oak, post oak

Adult Spiny Oak-Slug Moth.

Caterpillars of the Spiny Oak-Slug can vary in color.

Nanina Oak-Slug Moth   *Euclea nanina*   **4697.1**

This species should be easy to identify. Brown forewings with a large lime-green splotch are folded over the body. Short and stubby legs barely show from under the wings.

Tveten found this species at the Frio River cabins in April 2006 and 2007. In 2006, a female was captured and placed in a jar.

Eggs were laid a day after capture. A total of 80 to 90 clear, colorless eggs were laid in bunches on the lid of the jar. Tveten wrote that these were "almost as if (a) mess is covered with jellylike glue." He also noted that there were "surprisingly many and large eggs for so small a moth." The eggs started hatching eight days later.

Larvae were tiny ovals that were pale yellow. Each had tiny spines that could be seen under magnification. Since these larvae are known to sting, Tveten used a brush tip to place each on a post oak or willow oak leaf.

The larvae did not feed for two days. Then they started "slowly eating small depressions on both surfaces of post oak and willow oak." Within 13 days they were 4 mm long and green with orange tubercles.

Tveten noted that the larvae spun but failed to emerge.

Adult Nanina Oak-Slug Moth.

Larvae of Nanina Oak-Slug
raised by John Tveten.

Range in Texas: Tveten's specimen was found in the Texas Hill County. There is one public record of a larva found in Houston.

When Found in Texas: April in the Texas Hill Country. October in Houston

Food Plant: Oaks including willow, post, white

Saddleback Caterpillar Moth *Acharia stimulea* 4700

This small dark brown moth is compact. Short dark forewings are decorated with a few white markings. The area where forewings meet the body is fuzzy. Legs are also notably fuzzy.

The larva of this species will make one stop and take notice. The spiny brown caterpillar has a striking yellowish-green saddle. The saddle has a round hole in it on the top.

Tveten found two small larvae on a water oak in the Little Thicket. He noted that they appeared to be intermediate instars. Colors were pale tan-brown with a yellow saddle. As they grew, they turned a darker brown.

Each spun a small oval cocoon. Emergence was a month after pupating.

Range in Texas: Eastern Piney Woods from north of the Houston area

When Found in Texas: April to October

Food Plant: Water and willow oak

Adult Saddleback Caterpillar Moth.

Striking larva of the Saddleback Caterpillar Moth.

Small oval cocoon next to an adult Saddleback Caterpillar Moth.

Genista Broom Moth (name per BugGuide) *Uresiphita reversalis* 4992

The adult of this moth deserves to be noticed despite its small size. Long purplish-copper to reddish-brown wings form a slender triangle over the body. Random bits of black decorate the wings. Slender, threadlike, white foreleg and midleg are held away from the body.

Tveten found a mass of these larvae on whitestem, or wild indigo, on low dunes at Padre Island National Seashore. He wrote that there were "literally hundreds of these larvae, some on virtually every indigo plant. When tiny, they are in groups and skeletonizing the leaves. (When) larger, they are more solitary and consume the entire leaf. They do some minute webbing of the leaves. The top leaves of the plant are webbed and skeletonized to black-looking versions by the young."

Adult Genista Broom Moths showing variations in the wings between the female on the left and the male on the right.

Larva of the Genista Broom Moth.

Larvae are green, turning yellow-orange when mature, per Tveten. The larvae are covered with black and white raised markings with white hairs.

The larvae pupated in very flimsy silk cocoons seven to eight days when kept in a jar. Tveten noted that the "cocoons are so poorly made as to pull apart if removed."

Adults emerged seven to 13 days after pupating. Tveten noted that the adults have orange hindwings with dark rear borders. Because the adults sit in a triangular fashion, he was frustrated that he could not get the rear wings in his photos.

Female adults have grayer wings with paler bands on the forewing.

Tveten noted that adults of this species have long labial palps— "almost a snout."

Range in Texas: Throughout Texas except the Panhandle

When Found in Texas: March to November

Food Plant: Wild indigo, Texas mountain laurel, and bushpea

Tveten called this a "sod webworm." He wrote that it "sits with tightly rolled wings, so appears to be slender." The light tan moth has dark angular marks. This species has long, protruding labial palps. These scale-covered appendages on the face are part of the mouth and are used to sense if something is edible. The large black eyes stand out against the light body.

Eggs are tiny, whitish beige, and shaped like jellybeans. In captivity, the eggs were laid loose in the bottom of a jar and not stacked. Tveten suspected that the female drops eggs among grasses in the wild.

Eggs turned dark roughly 10 days after being laid and a pale larva with dark head and thorax emerged. Larvae fed slowly on St. Augustine grass and rustyseed paspalum. They skeletonized the grass blades and left a loose web with grass behind. Twenty days after emergence the larvae were beginning to turn brown with black heads. A month after hatching they were brownish gray and 14 mm or roughly ½ inch long.

The larvae chewed grass in the bottom of the jar and then pupated in the mess. Adults emerged within a few days of pupating.

Range in Texas: Throughout the state except the Panhandle and far west

When Found in Texas: All year

Food Plant: St. Augustine, rustyseed paspalum

Adult of the Profane Grass-Veneer.

Larva of a Profane Grass-Veneer on a blade of grass.

Indian Meal Moth *Plodia interpunctella* **6019**

We all hate it when this tiny ¼-inch-long moth flies out of a bag of bird seed. The adult flying free of the bag lets us know that the rest of the bag is infested with eggs, larvae, and pupae. Worse though is if the moths get free in the house our pantry gets infested as well. Every product in the pantry made with grain can be home to these annoying creatures.

Tveten found larvae in bags of birdseed. Caterpillars were a beige ivory color to whitish color. These create a moldy looking webbing in the seeds.

Cocoons were loose, flossy soft webs of silk. These were transparent enough to see the rusty tan pupae inside. Emergence was five to nine days after pupation.

Sunflower seeds infested with Indian Meal Moth adult and larvae.

Adult Indian Meal Moth.

One larva was observed to crawl into a loose cocoon and eat the pupa, leaving just fragments of the shell.

Range in Texas: Entire state

When Found in Texas: All year

Food Plant: Grain but also dried vegetables. Major pest.

Curve-Toothed Geometer *Eutrapela clemataria* **6966**

Throughout Tveten's notes about this species he uses the word "large." Wingspan in an adult can be 38 to 56 mm or 1½ to 2⅕ inches. Brown wings are held out from the body. The forewing is sharply hooked at the apex and hindwings each have a point along the outer margin.

One captive adult laid approximately 160 eggs in a jar overnight. The eggs were laid in short lines. The smooth, pale jade green eggs had no "sculpture" and were "just touching each other." The eggs were brown by the following day.

Tveten wrote that one larva found in Evergreen, Texas, was "enormous: 55 mm long and quite heavy." Another specimen was 40 mm long and 6 mm wide. Larvae have a tapering head, and wide rear, are gray-brown with prominent ridges, and have small tubercles. They look exactly like a twig and move in a traditional inchworm fashion.

Pupation was in a leaf stitched to the side of the jar. Another simply folded a leaf with no cocoon.

Adults emerged 13 days after pupating in April. Yet, a specimen in November emerged five months later in April. Another that pupated in April emerged four months later in August.

Adult Curve-Toothed Geometer showing variations in the wings.

Range in Texas: Tveten recorded these from the Piney Woods to his home in Baytown. Current records include to the Central Texas Coast and over to San Antonio.

When Found in Texas: All year

Food Plant: Willow oak

Larvae of the Curve-Toothed Geometer look like a twig and moves like an inchworm.

Large Maple Spanworm *Prochoerodes lineola* **6982**

The brown moth looks like a piece of a leaf with nearly perfect falcate wingtips. Both wings are held away from the body with the forewings forming an arch at the top. The hindwings each have a point that mimics a smooth leaf. Wings are brown with a fine dark line running from forewing tip to forewing tip.

Tveten found this moth around lights at his home in Baytown and in the Piney Woods.

An adult was put in a jar and it laid a few eggs a day later and approximately 50 eggs the next day. The eggs were "tiny blue ovals, like blue-gray jellybeans." Within two days the eggs had turned a reddish-purple and appeared to be slightly collapsed. Seven days later, the eggs hatched.

Large Maple Spanworm adults blend with a leaf.

Newly emerged larvae were very thin and long inchworms. Tveten wrote that they "move very rapidly with funny, jerky movements."

An older larva "looks like a twig, complete with leaf scars." It poses with its body rigid and held out from the branch when disturbed. Tveten called it a "prized camouflaged species."

The larvae were offered grass, willow, and willow oak. They ate voraciously on willow with some eating willow oak. Only those eating willow survived.

Cocoons were spun in leaves with adults emerging 20 days later.

Range in Texas: East Texas from Fort Worth to the Rio Grande Valley. Scattered records in Edwards and Val Verde counties.

When Found in Texas: March to October

Food Plant: Willow

Adult Horned Spanworm.

Horned Spanworm *Nematocampa resistaria* 7010

This angular geometrid holds all four wings out from the body. Each forewing and hindwing have an angle in the middle of the outer margin. There is a reddish-brown submarginal band on the hindwings but this only covers half of the forewing. Adults are otherwise beige with dark, thin veins.

The larva is a looper. It lacks midabdominal prolegs like most larva. It crawls forward with its back prolegs to form an arched back and then lifts the front part of the body and moves forward. Back prolegs are then lifted up and brought forward so the back is tightly arched, front legs are lifted and pushed forward.

Tveten found larvae of this species on post oaks in College Station. Adults were found at his porchlight in Baytown from April to June.

One of those adults laid seven eggs in a jar two days after being captured. The eggs were tan ovals that hatched in a few days.

Larvae only fed on willow oak despite being offered cedar elm and sugar hackberry. The dark brown loopers had prominent "horns" on the base of the thorax.

A month later the lone survivor pupated on a flimsy cocoon formed from a rolled leaf.

In another instance, a pupa was bare and "surrounded by a few strands of silk that attached several pieces of grass to the pupa." Tveten observed that the grass appeared deliberately collected for the purpose.

Larva of Horned Spanworm showing looper motion.

The adult emerged two weeks after pupating. Tveten noted that adult males had fewer red markings than females.

Range in Texas: Eastern Texas from Dallas to McAllen

When Found in Texas: April to May but August in the Rio Grande Valley

Food Plant: Willow oak

Larva showing "horns" on thorax.

Flimsy cocoon of a
Horned Spanworm.

Variation in wing patterns on adult Horned Spanworms.

Red-Bordered Emerald *Nemoria lixaria* **7033**

What a delight to wake-up to a pale green moth under the porchlight in the morning. Adults have pale green wings fringed with white. The fringe is bordered in red. Their abdomens are green with white spots bordered in red. Wings are spread and held flat against the wall or tree.

Tveten found red-bordered emeralds often under his porchlight at his home in Baytown.

Captured females laid 10 to 50 scattered eggs in a jar on the lid and vegetation inside the jar. The eggs were tan and flattened in an oval shape. As the eggs matured, they looked like "lozenges" and began to change to an orange color.

Larvae emerged 10 days after eggs were laid. The larvae emerged from a tiny hole in the tip of the egg. None ate the eggshell.

The larvae fed well on willow oak and eventually started skeletonizing the leaves. The larvae were brown with "flags" sticking out of the body segments. Unfortunately, all larvae in captivity eventually died.

Range in Texas: East Texas Piney Woods, to the Houston area, across to San Antonio and up to Fort Worth

When Found in Texas: All year

Food plants: Willow oak, but others report red oak

Variation in the wing patterns of Red-Bordered Emerald.

Eggs of the Red-Bordered Emerald change from tan to orange as they age.

Thin larvae of the Red-Bordered Emerald with "flag" projections.

Adult Southern Emerald.

Southern Emerald   *Synchlora frondaria*   **7059**

This is another showy green moth. Green wings are held open and flat. Each wing is decorated with filigrees of white. The green abdomen has a white line on top.

Tveten found these in the Baytown and Wallisville area from August to November.

While the adult moth is lovely the caterpillars are the showstopper of this species. The 10-mm larvae attach rough pieces of flowers along their back as camouflage. Tveten watched this "inchworm" feeding on a cowpen daisy and then attach pieces of the flower to its back.

In captivity, the larvae pupated on the side of the jar with a few strands of silk. Other specimens used silk ornamented with pieces of flowers as their cocoons.

Adults emerged two weeks to a month after pupating.

Range in Texas: Piney Woods to Rio Grande Valley over to the Hill Country with scattered records in the rest of the state

When Found in Texas: All year

Food Plants: Cowpen daisy

A Southern Emerald is hard to see because it has pieces of daisy attached to its body.

A pupa of a Southern Emerald rests inside a cowpen daisy bloom.

Packard's Wave *Cyclophora packardi* **7136**

This small, delicate tan geometrid moth holds both wings flat away from the body when resting. Evenly spaced dots outline the submarginal border of forewing and hindwing. Each wing has a small eyespot or stigma. Forewings are ½ inch long.

Eggs were laid like a tight string of pearls hanging from a leaf. Each string of eggs was on the edge of a leaf. Eggs hatched a week after being laid. Larvae fed on both post oak and willow oak with a preference for post oak.

At 20 days, the larvae were slender, lime green and 20 mm long.

Pupation took place about 25 days after eggs were laid. The green pupae were shaped like an elongated triangle with the base smaller than the sides. Each was attached like a butterfly pupa with a silken girdle.

Adults emerged seven days after pupating.

Range in Texas: Dallas and Austin area east to the Central Texas Coast plus Big Bend

When Found in Texas: All year

Food Plant: Post oak and willow oak

Adult Packard's Wave camouflaged on a leaf.

The Beggar *Eubaphe mendica* **7440**

This oddly named moth is pale yellow with almost transparent wings that are held flat away from the body. Forewings have grayish transparent cells.

Tveten found this moth under his porchlight in Baytown Texas.

He placed the adult in a jar and a day later she laid approximately 15 eggs. Each was a tiny, buff-yellow sphere.

The eggs hatched six days later. Tveten noted that the "tiny but long, slender loopers move really fast."

Violet is supposed to be the host plant for the beggar but Tveten could not find any wild plants. He tried maple but the larvae refused to feed.

Range in Texas: Eastern Piney Woods to Houston area

When Found in Texas: March to June

Food Plant: Violet and maple

Adult The Beggar moth.

Adults hold their light brown wings in a delta pattern to the side of the body. A wide band runs through each forewing and the base is curved. The shoulder area of the thorax is noticeably "hairy," and antennae are wide and fuzzy.

Caterpillars can be found literally by the handfuls in the right situations. Larvae have two noticeable blue lines that run the length of their body. The top dorsal surface has white dots against a brown background. The entire body is covered in short hairs.

Large larvae pupated nine days after Tveten placed them in a cage. Half of them spun a cocoon and half formed a bare pupa. Tveten described the pupa as "dark brown on dorsum; light brown ventrally, very active when disturbed."

During a heavy infestation at Palmetto State Park, Tveten noted that there were thousands of pupae on all kinds of plants. The cocoons were found on palmetto, oak, hackberry, poison ivy, box elder, buckeye, and "virtually anything." Box elder was a favorite where the larvae formed webs under a domed leaf. They "quit after there were two or three cocoons in a single webbed leaf," he noted.

Adults emerged seven days to a month after pupating in captivity.

Color variations in adult stage of the Forest Tent Caterpillar.

Detail of the body of a Forest Tent Caterpillar.

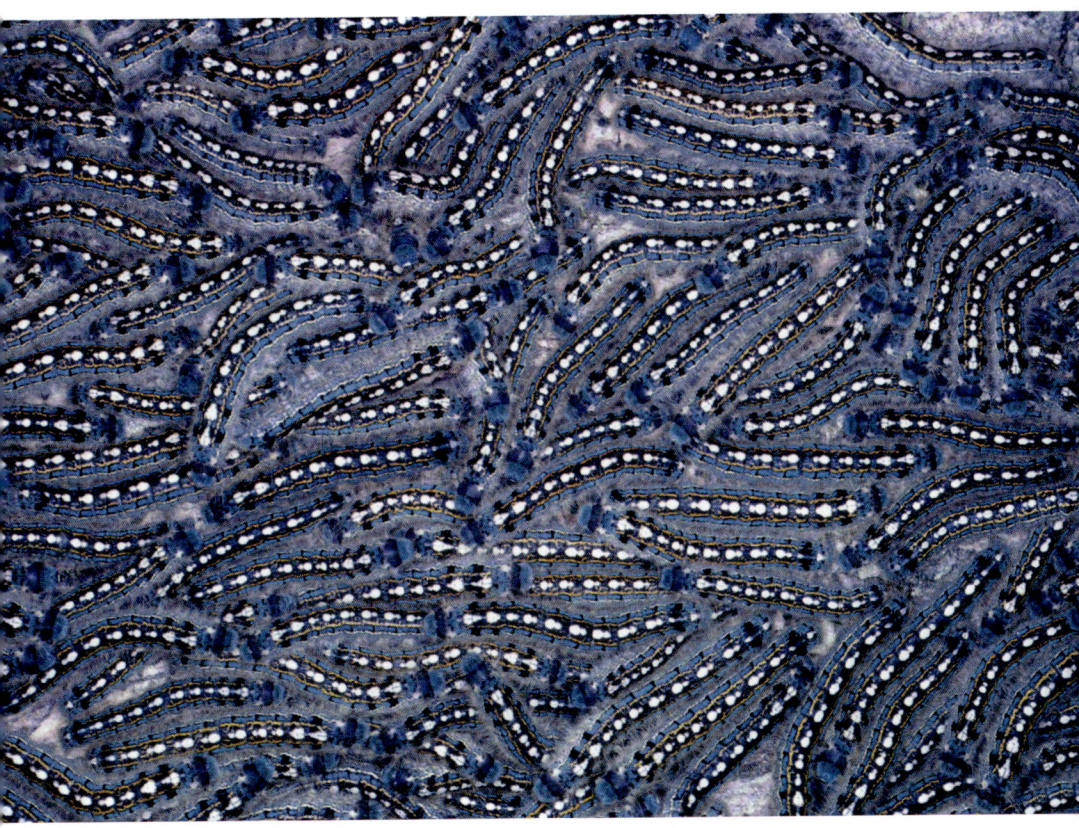

Mass of Forest Tent Caterpillars.

Range in Texas: Eastern Texas from Fort Worth south to San Antonio and over to Corpus Christi area

When Found in Texas: March to June

Food Plant: Live oak, willow oak, hickory

Eastern Tent Caterpillar *Malacosoma americana* **7701**

The scourge of homeowners and forest managers, tent caterpillars can do a lot of damage to vegetation.

Adults are burnt orange with a mass of hairs at the base of the forewings. Forewings are held over the body in a tentlike manner. Each wing has a lighter basal area and a lighter submarginal band. Antennae are large and feathery.

Tveten found a "tent" full of larvae along a road in Chambers County. The tent with about 200 larvae was in a hawthorn bush. Each larva was in the third to fifth instar and appeared to have wintered in the tent because the bush had only recently sprouted leaves.

Variations in the adult form of the Eastern Tent Caterpillar.

Eastern Tent Caterpillars live in a "tent" of webbing.

Detail of an Eastern Tent Caterpillar.

The tent was constructed of consecutive layers of silk with larvae between each pair of layers.

Several years later, Tveten found egg masses in a hawthorn bush in the same area. The eggs formed rings around the twigs. Each ring was black and layered with eggs.

Range in Texas: East Texas from Houston/Galveston to San Antonio/Austin and up to Dallas/Fort Worth

When Found in Texas: February to June. Only one generation per year in Texas.

Food Plant: Fruit trees and hawthorn

Western Tent Caterpillar   *Malacosoma californica*   **7702**

The Western Tent Caterpillar is very similar to a forest tent caterpillar with a thick body and wings held in a delta pattern. Adults have dark reddish-brown, to yellow, tan, or gray wings. Each wing has two lines that form a band. Wingspan is 1–2 inches. Antennae are "feathery."

Tveten found an outbreak at Guadalupe National Park in the area of Butterfield Stage "but there were none found elsewhere in the general area."

The eggs were hard, dark, and covered in a substance that looked like lacquer.

Larvae were found on oak and sumac. They refused to eat other species of oaks when offered.

Larvae that were moved into cages spun cocoons in three to five days.

Range in Texas: Scattered reports from Guadalupe State Park in West Texas to the Panhandle

When Found in Texas: April and May

Food Plant: Oak and sumac

Variations in adult form of Western Tent Caterpillars.

Detail of a Western Tent Caterpillar.

Imperial Moth *Eacles imperialis* 7704

The Imperial Moth can really grab your attention. Adults have a 4-inch wingspan. Males are purple like merlot wine with splotches of yellow. Females are more yellow with purple speckles.

Tveten found larvae in the Big Thicket, Little Thicket, and Big Creek Scenic Area of East Texas. He also found a larva in Houston's Eisenhower Park.

Larvae were described as "brown plain," "rich, rusty brown," and green. All larvae were found in October or November, so season did not seem to dictate color.

Imperial Moth larvae are large with slender "horns" on the head and tail. Segments T2 and T3 each have a horn. The longer horn is on the anal end. Shorter spines are on the remainder of segments along with tiny white hairs. When disturbed the larva will rear-up in a defensive posture.

In two instances, Tveten noted that after shedding skin the larvae were pale green.

Tveten found these larvae on a variety of plants including American sweetgum, loblolly pine, swamp chestnut, sumac, and redbud. Larvae feeding on red maple and red bay added new food plants to Tveten's

Adult form of the Imperial Moth.

Color variations in the larvae of the Imperial Moth.

Larva after shedding skin.

Interesting feeding behavior of the Imperial Moth larva noted by Tveten.

knowledge base. In one case, Tveten noted that a sweetgum was directly below the red bay but the larva was feeding on the bay.

Most of the larvae Tveten encountered were taken home to raise: "Tonight, while photographing I watched it eat a new leaf. First it crawled on the petiole and lay upside down and tried to pull back the leaf. Finding it resistant, it backed up a little ways, chewed halfway through the petiole in one place, and then promptly crawled back out and pulled down the leaf."

Larval stage ranged from one to five weeks.

Larvae pupated after burrowing into dirt. One "excreted much fluid" before burrowing. A few pupae were covered in fungus when uncovered or dried up due to parasites.

Range in Texas: East Texas from Dallas through the Piney Woods to Corpus Christi and into the Hill Country (iNaturalist)

When Found in Texas: April to November (iNaturalist)

Food Plant: Sweetgum, loblolly pine, swamp chestnut, sumac, redbud, red maple, and red bay

Royal Walnut Moth (a.k.a. Regal Moth) *Citheronia regalis* **7706**

Adult Royal Walnut Moths are gray, orange, and creamy white. The fuzzy body is orange with bold cream markings. Wings are gray with orange marks along the veins and punctuated with cream dots.

The larva in contrast is so ugly it is called a Hickory Horned Devil. Tveten made notes about seven encounters he had with this species. The specimens were found from the Little Thicket and Big Thicket areas to Houston.

All larvae were found in September and October feeding on sweetgum and one was found on a swamp chestnut oak. One larva was located by looking around an area where sweetgum leaves had been eaten.

The Hickory Horned Devils measured 5 inches long in the last instar. A young larva was 2 inches long and colored rusty brown with tan

Adult Royal Walnut Moth.

Young larva of a Royal Walnut
Moth with well developed
"horns."

patches. The "horns" were well developed. As this larva matured it turned
green but did not undergo a molt.

Larvae expelled a liquid before pupating. One shrank and turned a
darker shade of green before transforming into an "almost rigid cylin-
der." Tveten had dirt in the bottom of the screen enclosure and noted
that they all burrowed when ready to pupate.

The change from larva to pupa occurred 3 to 15 days after finding the
specimens. Pupae changed from iridescent green to rusty and finally
dark brown as each aged.

Later instar of the Royal Walnut Moth larva often called a Hickory Horned Devil.

Emergence times were not recorded. Tveten did note that one pupa was still buried in dirt two months after burrowing.

Range in Texas: East Texas Piney Woods to below Houston (iNaturalist)

When Found in Texas: April to October (iNaturalist)

Food Plant: Sweetgum and swamp chestnut oak but also pecan, walnut, and persimmon (*Butterflies & Moths*)

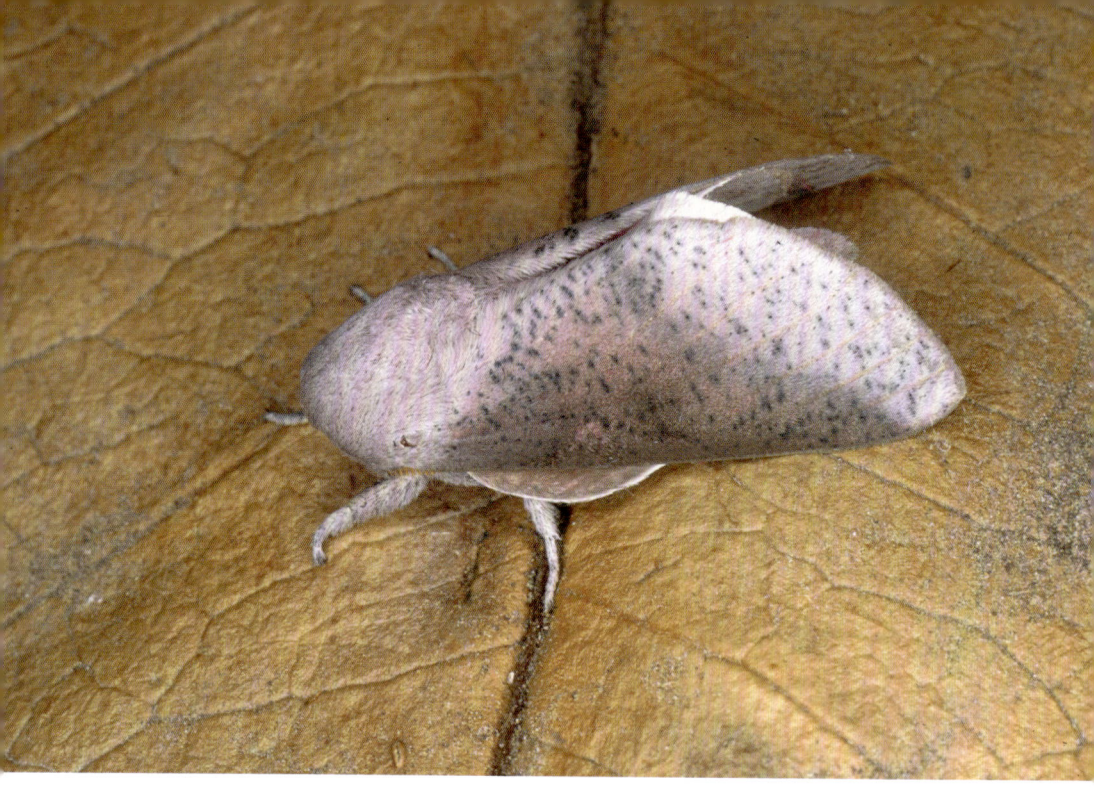

Honey Locust Moth   *Sphingicampa bicolor*   **7709**

This moth is dimorphic. The colorful form is more typical with orange forewings that have a faint brown band near the edge. Two tiny white dots decorate each wing. The less colorful form is solid gray and the two tiny white dots stand out. Wings are held tight to the body in a tentlike fashion.

Adult honey locust moths have a lovely rose hindwing. Tveten lamented that he could not get a photo of the moth with its hindwings exposed.

Tveten found this moth in Cameron County in deep South Texas. The adult came to a blacklight.

Range in Texas: East of Dallas and Austin and north of Houston

When Found in Texas: March to October

Food Plant: Honey locust

Color variations in the adult form of the Honey Locust Moth.

Heiligbrodt's Mesquite Moth *Syssphinx heiligbrodti* **7710**

Adults are similar to the Honey Locust Moth (*Sphingicampa bicolor*). Adults have gray forewings covering hindwings that are rosy at the base and gray toward the edge. A small black dot decorates the middle of the hindwing. The gray forewings have a lighter band across the middle that is edged in dark gray. Two tiny white dots barely touch near the costal edge.

Tveten wrote, "These are the most incredible larvae I have ever seen." He found them in October at Lake Corpus Christi State Park in San Patricio County. The green larvae had four long "horns" behind the head and another one at the posterior abdominal segment. There were "silver platelike spines on the sides."

Range in Texas: Central Texas primarily south of San Antonio

When Found in Texas: March to September

Food Plant: Black-brush acacia

Adult Heiligbrodt's Mesquite Moth with rosy hindwing.

Larva of Heiligbrodt's Mesquite Moth with four long "horns."

Hubbard's Small Silkmoth *Syssphinx hubbardi* 7711

This small, gray silkmoth rests with its forewings folded over more colorful hindwings. The hindwings are dusty rose at the base and gray along the edges. A small black dot decorates each hindwing.

Tveten found one larva in October on a catclaw acacia in Limpia Canyon in the Fort Davis area. The green larva was covered in horns with a lateral yellow stripe.

The larva became lethargic and pupated into a cocoon in tangled strands of silk 12 days later. The oval cocoon was 22 mm long.

Tveten wrote that the larva must have contained a parasite because the empty skin of the larva was outside the cocoon.

Range in Texas: Western Texas to central part of state

When Found in Texas: June to October

Food Plant: Catclaw acacia

Color variations in Hubbard's Small Silkmoth.

Larva of Hubbard's Small Silkmoth with "horns."

Rosy Maple Moth *Dryocampa rubicunda* 7715

There is no way to pass by a moth that is yellow with rosy markings. This moth is primarily pale yellow with a large rosy patch at the base of the forewing. A rosy band decorates the submarginal area of the forewing. The head and body are covered in dense yellow hairs.

Tveten found this species near Texarkana in the eastern reaches of the state and into the East Texas Piney Woods.

A mass of small larvae was found stripping the leaves from a red maple. All the larvae were clustered under a leaf. This activity was observed in July and again in October. Tveten determined that the larvae were in the second or third instar and measured 6 mm long.

A single freshly emerged adult was found in a tupelo tree in the Big Creek Scenic Area in April.

Adult Rosy Maple Moth.

Larvae clustered on the underside of a leaf.

Range in Texas: Eastern Piney Woods with scattered reports from Fort Worth and Kenedy County

When Found in Texas: March to October

Food Plant: Maples and oaks

Spiny Oakworm Moth *Anisota stigma* 7716

This is a lovely moth with deep orange wings covered in darker speckling. A dark postmedial line runs from the apex of the forewing diagonally to the inner side. The hindwing is covered when the adult is roosting.

Tveten encountered this species in the Piney Woods of East Texas between August and October.

In one instance he noted that they are "very common along with several other oakworms." Years later he wrote, "Oakworms are prolific this year for the first time in four or five years."

In the Little Thicket, Tveten found larvae on willow oak, laurel oak, water oak, red oak, and white oak. The larvae had stripped very small oaks of their leaves.

Larvae were 6–7 mm long with a large orange head. The body was blackish with a dorsolateral stripe. Each segment had spines. In a later instar, a large larva had wide pink and green stripes and white dots. Tveten noted that the skin appeared dry and the spines did not sting. The largest larva was 60 mm long and 10 mm thick.

Pupae were bare and lay in the open on the bottom of their container. Some pupae were eventually covered with a white mold and died.

Adult Spiny Oakworm Moth.

Larva of Spiny Oakworm Moth.

Female adult Spiny Oakworm Moth.

Seven months after pupating, one adult female emerged. Tveten waited one more year but no other adults emerged from their pupae.

Range in Texas: Eastern Piney Woods to the Austin and San Antonio area

When Found in Texas: June to October

Food Plant: Oaks including willow, laurel, water, red, and white

Orange-Tipped Oakworm Moth *Anisota senatoria* 7719

Males of this species have red translucent forewings that are sometimes speckled. Females have orange forewings with light speckling. Both have a white spot on the forewing.

Tveten found these in East Texas from the Kirby State Forest to the Little Thicket. One October, he noted that there were "tremendous numbers on lawns." Along with other oakworm species they were stripping many of the small trees. He found them on various oaks, including water and willow oak.

A "leaf full of eggs" was located on a water oak. Some had hatched and Tveten watched as others hatched.

The tiny larvae were pale green with black heads. Each had three faint dark dorsal stripes and a pair of dark spines behind the head.

Larvae brought into captivity had the same appearance six days later. Seventeen days later they were still growing and confirmed as orange-striped oakworms (the former name.) One mature larva measured 45 mm long.

Another batch of larva in the wild had an orange base color rather than black and the skin was shiny.

Pupation yielded a bare pupa in the bottom of the container.

Range in Texas: East Texas Piney Woods with report in the Dallas area

When Found in Texas: January to October

Food Plant: Oaks including water and willow

Larva of the Orange-Tipped Oakworm moth.

Pink-Striped Oakworm Moth *Anisota virginiensis* **7723**

At rest, the Pink-Striped Oakworm Moth can be 1 inch across. Dark orange-colored forewings have a dark diagonal postmedial stripe. The leading edge of the wing is a bit darker than the trailing edge. One white spot decorates each forewing. Tveten described a male as having triangular wings with "windows."

Tveten found this species several times from the East Texas Piney Woods to the Houston area. Larvae were found from May to October on basket oak, water oak, and red oak. A larva found on basket oak fed on post oak when offered.

Small larvae were yellowish orange with short black spines and measured approximately 8 mm long. A pair of larger spines were on the second thoracic segment. The spines did not have barbs on the ends. These measured 8 mm long.

Larger larvae were pinker and greener with an orange-green head. The body spines were very short.

Pupation began a month after the small larvae were placed in an enclosure.

Adult Pink-Striped Oakworm Moth.

Larva with barbs on oak leaves.

    Adults emerged three weeks after pupating in the summer. One adult emerged after only six days in a pupa. Another that pupated in October emerged the following May.

Older larva of the Pink-Striped Oakworm Moth showing the orange-green head.

Range in Texas: East Texas from Dallas area to Austin and Houston

When Found in Texas: April to October

Food Plant: Oaks including basket, water, and red

Adult Eastern Buck Moth.

Eastern Buck Moth *Hemileuca maia* 7730

The adult stage of a buck moth is black with a white band that runs through the center of each wing. A small eyespot decorates each wing. Adult abdomens are fuzzy and tipped in orange.

Tveten found larvae at Inks Lake State Park in Burnet County and in Uvalde County in the Texas Hill Country. Each encounter was in April.

Larvae were found on live oak and red oak trees.

This species is one of the stinging caterpillars. Tveten noted, "Stung my hand badly. Some spines broke under skin. Very bad—lasted couple of days."

The larvae pupated in a cage. They emerged six months later in October.

The spines on Eastern Buck Moth caterpillars can deliver a nasty sting.

Range in Texas: Eastern part of the state from Dallas through San Antonio area and down to Corpus Christi

When Found in Texas: All year

Food Plant: Oaks

Grote's Buck Moth (iNaturalist calls this Buckmoth)
Hemileuca grotei 7733

This moth is very similar in appearance to the Eastern Buck Moth. Wings are black with a wide white stripe cutting through the center. Tiny black eyespots decorate each white band.

Tveten found several larvae in April 1992 in McKittrick Canyon in the Guadalupe Mountains. The dark spiny larvae were on the branches of gray oak. Each larva was approximately 1¼ inches long and black with a pattern of small spots.

The larvae molted in 10 days and were 35 mm, or 1.37 inches long. Each was black with red rings and yellow lateral stripes. Spines covered each body. Tveten noted that these spines did not sting.

Adults emerged in late September five months after the larvae were discovered.

Range in Texas: Central Texas

When Found in Texas: All year

Food Plant: Live oak

Adult Grote's Buck Moth.

Eggs of a Grote's Buck Moth laid on a live oak twig.

Larva of the Grote's Buck Moth.

Saturniidae

(Copied from Tveten's November 22, 1996 article in the *Houston Chronicle*)

Although butterflies have achieved widespread popularity in recent years, their nocturnal counterparts, the moths, have failed to gain similar acceptance. Moths, however, far outnumber butterflies and exhibit astounding beauty and diversity. Some rank as the most majestic of all our insects.

Chief among these masters of the air are the giant silkmoths in the family *Saturniidae*. Some, like the Atlas and Hercules moths, are enormous, spanning as much as 12 inches. Others sail gracefully through tropical rain forests with amazingly long, slender tails trailing far behind.

Born without functional mouth parts or digestive systems, silkmoths are unable to eat or drink as do butterflies and most other moths. Each lives but a few days, its entire adult life dedicated to mating and laying fertile eggs to perpetuate the species.

In spite of their size and beauty, these saturniids have received little attention from artists and authors through the ages.

The family *Saturniidae* contains more than 1,500 species, the majority inhabiting remote tropical forests. Fewer than 100 occur in North America, but these include such well-known moths as the luna, cecropia, Polyphemus, Promethea, imperial, royal walnut, and Io.

Adult Promethea Moth.

Adult Promethea Moth in different color.

Underside of a Polyphemus Moth resting on an oak leaf.

Io Moth *Automeris io* 7746

Male Io Moths have lovely golden wings that are smudged in purple. The forewings are held so they cover two huge eyespots on the hind wings. Each eyespot is ringed in black and purple so each eye looks like that of an owl. Females have wine-colored forewings that cover the same eerie eyespots. Females can also be dark grayish brown like a piece of weathered wood.

Tveten found eggs, larvae, and adults from the East Texas Piney Woods, to the Hill County, and down to the Rio Grande Valley.

He was able to get pairs to mate in captivity and lay eggs. One pair yielded 130 eggs and another mating resulted in 181 eggs. Mating took place within a day of the female emerging and eggs were produced two days later.

In Bracketville, Tveten found 30 Io eggs on a blade of grass. Each was white and shaped like a kernel of corn with a black dot on top.

Eggs hatched around two weeks after being laid. Larvae were spiny and greenish-orange in color.

Larvae tended to stay together and behave as a group. They "spent one full day at the hatching spot, ringed around eggs, ate shells completely." After that all the larvae moved off the leaves in single file, end-to-end.

Color variations in the Io Moth.

The second instar was reached after a week and the larvae were still feeding as a group.

By the third and fourth instar, the larvae were much bigger and darker but still greenish-orange with a longitudinal white stripe. Each was

A female Io Moth laid eggs within a day of emerging.

Detail of an Io Moth larva.

Some of the spines on an Io Moth larva are black and white.

ringed with tubercles with brushy spears. The spears, or spines, were black and white.

Tveten noted that larvae were approximately ½ to 2 inches when grown. Two larvae raised in October were noted to be "fairly small" with resulting small pupae.

Food plants varied. Tveten fed larvae mesquite, willow oak, Chinese wisteria, post oak, laurel oak, basket oak, white oak, water oak, and St. Augustine grass. Tveten found individuals on sugar hackberry, eastern red cedar, sassafras, flowering dogwood, and Carolina buckthorn. These were all new host plants for him.

The pupa stage was spent in a "very small" cocoon. Cocoons were spun in leaves and even in the bottom of a jar.

Io Moths often hold their wings so the lovely eyespots are covered.

Emergence took place one to two months after pupation. Larvae that pupated in late fall emerged three to six months later. Yet, a few adults raised inside emerged in December and January. Adult females were found around porch lights in February in the Rio Grande Valley.

In one case Tveten noted that adults emerged between 6:00 P.M. and 9:00 P.M. over a series of days.

Io larvae can sting with the spines on their body.

Range in Texas: Eastern half of the state from Dallas down to Laredo and eastward plus the Davis Mountains area

When Found in Texas: February to November

Food Plant: Mesquite, willow oak, Chinese wisteria, post oak, laurel oak, basket oak, white oak, water oak, St. Augustine grass, sugar hackberry, eastern red cedar, sassafras, flowering dogwood, Carolina buckthorn

Polyphemus Moth   *Antheraea polyphemus*   **7757**

Polyphemus Moths are a showstopper. Adults have large tan wings with blackish eyespots on each hind wing. The wingspan can range from 3 to 5 inches.

Tveten found this species from Coldspring in the Piney Woods through the Houston metropolitan area.

He found a freshly emerged adult female in August at the base of an oak tree in his backyard in Baytown, Texas. Twice he was given Polyphemus eggs from a friend in Houston in October.

Polyphemus larvae have pale green segments with vertical pale yellow slashes. Those slashes are dotted with red spiracles (openings where the worm breathes). There are three yellow projections, or "bumps," on each segment. Each bump has three or four hairs.

Tveten noted one large larva was 50 mm long and 15 mm thick. (1.96 by 0.59 inches)

Polyphemus larvae feed on willow oak and post oak leaves. These leaves are also used when the larvae mature and are ready to spin a cocoon. The white oval-shaped cocoons are about 1½ inches by 1 inch and wrapped between two leaves.

Adult Polyphemus Moth.

Eggs of a Polyphemus Moth on a tree branch.

Polyphemus Moth larva.

The cocoon of a Polyphemus Moth can be wrapped in leaves.

Emergence from the cocoon occurs about two months later.

Range in Texas: Eastern Texas from Dallas to San Angelo through San Antonio to Aransas Pass

When Found in Texas: February to December

Food Plant: Oak, willow, maple, and birch

Luna Moth *Actias luna* **7758**

The lovely lime-green Luna Moth will get anyone's attention. Tveten documented his encounters with this species from 1978–98. Those encounters were all in the East Texas Piney Woods primarily in the Little Thicket and Big Creek Scenic Area.

Larvae were found in October and November and once in late April. Larvae were found on sweetgum trees. A search of nearby sweetgum trees often resulted in finding more larvae.

Adult Luna Moth.

Variations in the spotting on Luna Moth larvae.

Luna Moth larvae are green and Tveten noted their size from ½ inch to 2 inches long depending on their stage of development. Each segment of the larva has protruding red spots called tubercles. The size of the red spots varies per individual and one specimen had no spots at all. Tveten commented that one larva had very large red spots on the body. He wrote that this was "the prettiest luna larva I have seen."

Dark spots were noted on some larvae. Tveten noted the dark spot was a suspected parasite. In the fall of 1993, about 40 parasite larvae emerged from one caterpillar Tveten was raising in a screen cage. The parasites spun small white cocoons upon emerging.

The luna larvae fed on sweetgum leaves. Tveten noted that in October 1986 many sweetgum trees in the area had only the leaf petioles remaining. Petioles connect the leaves to the stem of a tree so the larvae had fed only on the leaves.

Sweetgum leaves were used again as the larva transformed into a cocoon. Tveten's notes show that metamorphosis started about 10 days after larvae were collected. Larvae spun in the leaves on the bottom of their enclosure. One larva formed a bare pupa that eventually was covered in a white fungus.

Luna Moth pupae can be active. Tveten noted that several "jump when handled."

Emergence from the cocoon occurred 24 to 51 days after spinning. One specimen spun into a cocoon in October and emerged the following March or five months later.

Range in Texas: Fairly common from Dallas down to San Antonio and east to the coast and Piney Woods

When Found in Texas: All year

Food Plant: Sweetgum, persimmon, hickory, walnut, sumac

Forbes' Silkmoth   *Rothschildia lebeau forbesi*   **7761**

Forbes' Silkmoth is a very large silkworm moth with a wingspan of 10 cm, or 4 inches. There is a clear "window" on each wing hence the Spanish name "cuatro espejos" or four mirrors.

In May of 1989, Tveten found a cocoon of this moth while camping at Bentsen-Rio Grande State Park, near Mission, Texas. The cocoon was placed in a screen cage and left on a shelf in Tveten's garage.

In April 1992, Tveten wrote, "A gorgeous *Rothschildia* female in perfect shape emerged in the screen cage. After three years!"

On that same camping trip in May of 1989, Tveten found a cocoon of a similar species with the Latin name *Rothschildia cincta*. That cocoon was drop-shaped and hanging from a dead branch. The cocoon "rattled" when touched to scare predators. Tveten left this cocoon to mature in nature and noted that there were other empty cocoons nearby.

Range in Texas: Rio Grande Valley

When Found in Texas: Cocoons February to July. Adults June to October.

Food Plant: Lime prickly ash, Mexican ash, and willow

Cocoon of the Forbes' Silkmoth.

Larva of the Forbes' Silkmoth.

Female Calleta Silkmoth.

Calleta Silkmoth *Eupackardia calleta* 7763

People in the moth and butterfly community frequently share larvae. Female moths can lay hundreds of eggs and someone can have hundreds of hungry worms when those eggs hatch. The best way to ensure all those larvae reach adulthood is to share them with others in the community.

Tveten received three large Calleta Silkmoth worms from a friend on October 19, 1984. The friend had determined that the larvae would eat privet but not willow.

The larvae fed sparingly on the small leaves of Chinese privet. They would not touch Japanese privet that have larger leaves. Tveten also had success feeding the larvae the leaves of ash trees.

Molting began five days later with the smallest of this group. This larva was now yellow with blue and black spines. The other two larvae molted shortly thereafter with Tveten noting that they were "now incredibly beautiful." He wrote that they were "pale aqua with orange and black spots. Orange is around base of each tubercle. Tubercles ringed at neck with black with rounded turquoise knobs and slack spines. Claspers yellow with black border and spots."

Many years later, in the winter of 2002 Tveten found several cocoons on cenizo at the Inn at Chachalaca Bend in Cameron County. He assumed that they were spun in the fall of 2001.

These cocoons sat on a shelf in Tveten's garage and started emerging in July 2003. The first to emerge was a female that he described as a "large,

Male Calleta Silkmoth.

Larva with a colorful body.

Larva of a Calleta Silkmoth beginning to spin a cocoon.

Cocoon of
Calleta
Silkmoth.

Eggs of the Calleta Silkmoth.

lovely, very dark moth." The cocoons that were found a year and a half earlier were viable and producing adult moths.

Very pale turquoise-white oval eggs were laid nine days later. Females laid more eggs, but we have no notes to tell us how those eggs fared.

One lone cocoon from the original 2001 batch sat on Tveten's shelf until September 2003 when a male Calleta Silkmoth emerged "presumably almost three years after pupating."

Range in Texas: South Texas north to Austin and west to the Davis Mountains and Big Bend

When Found in Texas: All year

Food Plant: Ash, cenizo, ocotillo, privet

Male Promethea Moth.

Promethea Moth *Callosamia promethea* **7764**

In Greek mythology, Prometheus created humans from clay and eventually gave them fire. The female Promethea Moth is clay-colored with a 2-inch wingspan. On the upper surface, her wings are deep russet-brown toward the body with wide light rufous borders. Each forewing has a prominent eyespot. Males have solid black wings edged in golden yellow with each eyespot edged in rufous.

Tveten wrote notes about three encounters with Promethea larvae.

In November of 1978 he wrote, "Found larvae in Big Creek Scenic Area. May also have another cocoon. This is the first evidence of Promethea I have ever seen in Texas in 18 years—either larvae, pupa, or adult."

The larva Tveten found that day began spinning in a leaf before he got it home. The larva was covered with a white waxy secretion. The next day the cocoon covered in golden silk was complete inside a folded leaf. The silk was fastened to the midrib of the leaf, the petiole, and the twig so that the "leaf cannot fall in winter."

Promethea larvae can be hard to spot, according to Tveten. Larvae are green with "horns" on the head. The third thoracic segment is raised. Prolegs are on abdominal segments three, four, five, and six. The end of the abdomen is lifted and elongated. Dark brown triangles are on the sides and the body is covered with tiny, light-colored tubercles.

Tveten found his larvae on arrowwood viburnum and sweetleaf. These larvae were located by following the stripped leaves. The larvae ranged from 25 to 50 mm, or 1 to 2 inches.

Promethea Moth freshly emerged from its cocoon.

Different stages of Promethea Moth larvae.

A covering of a white powdery secretion signaled the beginning of pupation. The larvae spun leaves together to form a cocoon. One was a bare pupa with no leaves.

Adults emerged 13 days after spinning in the summer.

Range in Texas: Dallas to the Piney Woods above Houston

When Found in Texas: March to November

Food Plant: Arrowwood viburnum, sweetleaf, spicebush, sweetbay, and sassafras

Promethea Moth larva.

Cecropia Moth   *Hyalophora cecropia*   **7767**

"I thought the enormous cecropia was the most amazing creature I had ever seen," said Tveten, describing his experience as a young boy in Minnesota when he first saw the moth. "For that, I owe the species a debt of gratitude, for my love of moths still endures today" (copied from a column by Gary Clark in the *Houston Chronicle*).

Friends and acquaintances knew that Tveten was raising moths. Tveten noted many times in his field notes that someone gave him larvae, cocoons, or adults.

In the spring of 2006, staff at Jesse H. Jones Park and Nature Center in Humble, Texas, gave Tveten four Cecropia moth cocoons. Tveten was 30 years into his study of moths and he wrote that the gift was "apparently not recorded on these records."

An adult Cecropia emerged from each cocoon in early May 2006. Tveten noted that one was deformed.

Cecropia moths and other silk moths are in decline. Reasons are loss of woodland habitat, increased use of pesticides and death from imported parasitoid flies (*Compsilura concinnata*) used to kill Gypsy moths.

Adult Cecropia Moth.

Cecropia Moth eggs on a twig.

Range in Texas: Hill Country east to the Central Coast and Piney Woods. Records in the eastern Panhandle.

When Found in Texas: February to September

Food Plant: Maple and other trees including alder, apple, ash, beech, birch, cherry, dogwood, elm, plum, white oak, and willow

Early instar of Cecropia Moth larva.

Later instar of Cecropia Moth larva.

Cecropia Moth cocoon on a twig.

Sphingidae

(Copied from Tveten's January 10, 1997 column in the *Houston Chronicle*)

Some of the moths are small and drab, attracting little attention as they swirl through the night skies. Many, however, are patterned in brilliant colors, while still others rank among the largest and most beautiful of all our insects. Some of the most striking are the sphinx moths of the family *Sphingidae*.

Sometimes called "hawk-moths," sphinxes have tapered, streamlined bodies and long, narrow wings. The worldwide family contains approximately 1,000 species, with wing spans as wide as 8 inches. Although the majority are tropical, many also are found across the United States and in Texas.

Among the fastest of all Lepidoptera, sphinx moths have been clocked at up to 30 miles per hour. Their powerful wings may beat as many as 40 times a second as they hover to sip nectar from flowers. In flight, they strongly resemble hummingbirds and, indeed, are often mistaken for those smallest members of the bird world.

Various sphinx moths use many of the same flowers visited by hummingbirds. Some are active at night; some, almost exclusively by day. Many, however, are crepuscular, feeding primarily in the fading light of dusk and again shortly near dawn.

The sphinx probes the deep calyx of a flower with a long proboscis, or "tongue," normally carried tightly coiled beneath its head. Unfurled, this proboscis may be longer than the insect's body, and is often thought to be the "beak" of a supposed hummingbird.

A few tropical species have a proboscis nearly a foot long, with which they probe the long tubes of orchids and other deep-throated flowers.

Tersa Sphinx "hawk-moth."

Pink-Spotted Hawk Moth *Agrius cingulata* **7771**

People often mistake this moth for a hummingbird. It hovers while feeding on nectar-rich flowers and is an important pollinator. Prominent pink areas are revealed on the abdomen in flight or when the wings are spread. When the wings are closed, the adult is mottled brown like tree bark or leaf litter.

Tveten found an adult under his Baytown porchlight in October 1989. He noted that it had a "buzzy flight" and that its spiny feet stuck when crawling on his hand.

Range in Texas: Entire state except El Paso area

When Found in Texas: May to October

Food Plant: Sweet potato, jimsonweed, pawpaw

Adult Pink-Spotted Hawk Moth.

Tobacco Hornworm (Carolina Sphinx) *Manduca sexta* 7775

The Tobacco Hornworm moth can be mistaken for a hummingbird as it feeds on flowers. Adults are large with mottled brown wings. In flight, the abdomen is sharply tapered at the end with yellowish dots on the side.

Larvae are often found on tomato plants. Tveten found larvae in early April 2000 in Cameron County. He noted that a larva was devouring a tomato plant in Livingston in July 2001. That large larva was green with white slashes on the sides. The slashes ran diagonally along the abdominal segments.

An adult was found in Real County along the Frio River under a blacklight in April 2007.

Range in Texas: Entire state

When Found in Texas: March to November

Food Plant: Tomato, potato, tobacco, pepper, jimsonweed

Tobacco Hornworm adult.

Larva of the Tobacco Hornworm.

Pawpaw Sphinx *Dolba hyloeus* 7784

Adults of this moth are triangle-shaped with mottled black and white wings. The body is thick with an abdomen that tapers to a blunt tip. Wings are held in a triangle alongside the body.

Eggs are transparent, light green ovals. "Size of a blunt pencil tip," per Tveten. Eggs are scattered and not deposited in an orderly fashion. One adult female in captivity laid seven eggs one day and 80–100 eggs the following day.

Six days later the eggs hatched. The first instar caterpillars were pale green with a black horn on the tail. The caterpillars had light green diagonal lines on the side.

Tveten found larvae in September 1981 in Tyler County and in October 1986 in San Jacinto County.

The larva found in 1981 was medium-sized, size, 37 mm long, and moderately slender. Tveten described it as "rich, velvety black with charcoal-brown on dorsal surface (shades into black on sides). Horn— long, black, pebbled. Wide white slashes above V in front of spiracles on abdominal segments 2–7. Mark on #7 running up into base of horn. Spiracles white with black center."

Mating pair of Pawpaw Sphinx.

Green larva of the
Pawpaw Sphinx.

A week later, this larva was feeding well on yaupon. Its dorsal surface was becoming "brighter in color."

Pupation began 10 days after capture. The larva was still the same color. It burrowed into peat under sphagnum. The pupa was brown with a short "tongue area" or "handle" that contained the proboscis.

An adult emerged 10 days later. Tveten wrote that the larva must have been melanistic since it was not the typical green larva shown in most books.

Pupae from another group formed in soil or on top of soil. Tveten wrote that, "Pupa quickly turns color: lime green to light brown to dark brown." These emerged two weeks after pupating per Tveten's notes. Two moths paired immediately after emerging with the female laying eggs one day later.

Tveten found two larvae in the Little Thicket area in October 1986. He noted that these were found on yaupon. Both were left in the field and he wrote, "These are the first I've found in several years."

Range in Texas: East Texas from Dallas to San Antonio and eastward

When Found in Texas: All year

Food Plant: Yaupon, blueberry, pawpaw, sweetfern

Dark charcoal-brown larva of the Pawpaw Sphinx.

Larva with parasite cocoons attached.

Catalpa Sphinx *Ceratomia catalpae* 7789

This brown nondescript hummingbird mimic has inch-long forewings. A closer examination shows brownish and gray patterning of lines. The thick tapered body has markings along the side of the abdomen but no prominent dorsal marks that can be seen in flight.

Tveten spotted a badly riddled catalpa tree along the road near Rusk, Texas. He stopped and collected 10 larvae that were within reach. He noted that there were many higher in the tree. The larvae ranged in color from light to entirely dark on the dorsum and they stayed on the underside of the large leaves.

Larvae at 70 mm long had a black base with yellow sides.

Pupation in captivity was either on top of the dirt or in a burrow.

Adults emerged 16 to 21 days after pupating.

Range in Texas: East Texas from Dallas to Houston. Report in Rio Grande Valley.

When Found in Texas: June to September

Food Plant: Catalpa

Adult Catalpa Sphinx.

Catalpa Sphinx larva feeding on a catalpa leaf.

Larva of the Walnut Sphinx
showing the "horn" on the head.

Walnut Sphinx   *Amorpha juglandis*   7827

The adult Walnut Sphinx has walnut-colored wings that could mimic tree bark. Lighter tones run horizontally when the wings are folded. Edges of the forewing are wavy. The small hindwing shows near the top of the forewing when the moth is at rest.

Tveten encountered an adult of this species under the porch light at his Baytown home in May of 1985. He noted that the moth was reddish. He found other adults attracted to a blacklight in Concan, Texas, in April of 2006, 2007, and 2009.

He placed a 2009 specimen in a jar with food to see if she would lay eggs. No luck but the following night with new pecan leaves, the moth laid eggs on the jar and leaves. Four days later the female was still laying eggs.

Eggs were green ovals with no ornamentation.

The eggs started hatching eight days after being laid. The small larvae fed along the edges of fresh pecan leaves. Eight days later they were feeding on larger pecan leaves. A "horn" developed on each larva's head. Fourteen days after hatching the larvae were mid-instar. Twenty days after hatching the "horn" was gone and the head capsule was pointed on top. Tveten noted that each had a "surprisingly small head."

Pupation began a month after the eggs were laid. Peat moss was placed in the bottom of the container and accepted. All the larvae had pupated on top of the peat moss within a week.

Pupae were "dark brown, hard, no proboscis tube."

Range in Texas: All of Texas except Rio Grande Valley and arid border with New Mexico

Color variations in Walnut Sphinx moths.

When Found in Texas: March to October

Food Plant: Pecan with black walnut, cherry, chestnut, hickory, and hophornbeam (in the literature)

Vine Sphinx *Eumorpha vitis* 7864

This moth has a strong triangular shape with dramatic tan bands on dark wings. The abdomen is thick with a sharp point at the tail end. A small underwing has a tiny bit of salmon color.

Tveten was given a huge sphinx larva found outside the Houston Museum of Natural Science in September 1981. He described the larva as very large at 95 mm by 17 mm. It was green and lightly sprinkled with small black dots with a light yellow-green ring around the spiracles. It had light diagonal stripes through the spiracles on latter abdominal segments. This larva had no horn. It had a habit of "withdrawing head and first two thoracic segments deep into body."

Larval colors range from pink to green in the literature. Tveten found Vine Sphinx larvae three times from Houston to the Rio Grande Valley and all were green.

The Houston larva dug in the dirt and was ready to pupate one day after it was found. It crawled into sphagnum to pupate. The Rio Grande Valley larva buried itself in dirt.

Tveten described the pupa as "very long and brown" with no tongue cord.

Adult Vine Sphinx moth.

All the larva of Vine Sphinx that Tveten found in Texas were green.

One adult emerged three weeks after pupating. Another found in late October emerged three and a half months later in early February.

Range in Texas: Entire state except Panhandle

When Found in Texas: February to November

Food Plant: Observed feeding on marine ivy. Literature also notes feeding on grape leaves.

Adult of the
Banded Sphinx.

Banded Sphinx or Lesser Vine Sphinx *Eumorpha fasciatus* 7865

This moth really lives up to its name. The adult is shaped like a typical, triangular sphinx moth. Tan bands edge the wings with more broad bands covering the wings. These bands are echoed on the body. A smaller hindwing shows rose color and a faint eyespot. The adult's abdomen is sharply tapered at the end.

Tveten found a larva in August 1986 at Hyatt Lake in Tyler County. The colorful larva was feeding on bushy, low-growing plants in a roadside ditch filled with pitcher plants. It was 60 mm long and heavy. He described it as follows: "Only blunt projections at anal end of back. Apple green below with red prolegs. Yellow, red, white, black above. Black dorsal stripe, flanked by yellow stripes. Long white dash across each segment behind dark spiracle."

Another group of larvae was found in July 2005 around a backyard pond in central Houston. This group was composed of specimens showing all the stages of the last instar.

Banded sphinx larvae change with each instar. During the last instar we see a large green caterpillar with a black dot on the side of each segment. Each dot is surrounded by an elongated spot of white.

Larva of showing color pattern in stage of development.

The large caterpillar burrows into dirt when it is ready to pupate. A caterpillar that pupated on July 23 emerged nearly 20 days later on August 12.

Range in Texas: Eastern Texas from Dallas to Austin and the upper Texas Coast. Scattered reports from Del Rio to Odessa.

When Found in Texas: April to November

Food Plant: Evening primrose

Adult Lettered Sphinx moths are mottled brown with jagged edges to the forewings. The abdomen is tapered but not smoothly tapered like other sphinx moths. Larvae are greenish yellow with short, light dorsal stripes.

Tveten found many larvae at the Houston Arboretum in April 1979 on native grape and Virginia creeper. He took several larvae home and within 15 days they were trying to pupate in dry dirt at the bottom of the cage. Each dried up and shriveled. Tveten's son, Michael, had success getting some larvae to pupate in moist dirt. Moist dirt was the key to survival.

Three years later, in April 1982, Tveten again found larvae on Virginia creeper at the Houston Arboretum. Each was green and yellowish.

These larvae were moved to captivity and given peppervine as a food plant. They easily adapted to the new food and ate well.

The larvae turned pinkish-green two weeks after being discovered. Tveten noted that they appeared ready to pupate. Each burrowed beneath moist sphagnum moss. The pupae were well formed several days later with sphagnum sticking to the outer shell.

Another group of larvae were found at the Little Thicket in San Jacinto County in May 1988. This group was feeding on the greenery of muscadine grape vines. These larvae were 32 mm long. Each was lime green with a long "horn" at the tail end. Tveten noted that each had "yellow rows of dots and an oblique line" along the abdomen with a narrow head and thorax.

When disturbed the larva "curls back with legs in air."

Range in Texas: East Texas from Dallas to San Antonio and eastward

When Found in Texas: February to April

Food Plant: Muscadine grape, Virginia creeper, wild grape, peppervine

Larva of the Lettered Sphinx showing green and yellowish color.

Tersa Sphinx *Xylophanes tersa* 7890

This moth is shaped like a stealth fighter jet. Its long, tan swept-back wings are sleek with clean edges. At rest the wings are held out from the body at an angle. The brown body is thick and sharply tapered at the end of the abdomen.

Tveten found specimens in the fall from the Kirby State Forest in Tyler County, to Kountze in the Piney Woods, to Conroe and Baytown in the Houston area, and down to Mission in the Rio Grande Valley.

Larvae were usually located alone but once with another nearby. Food plants were forked bluecurls and buttonweed in the wild, and penta in backyard gardens.

One larva was 35 mm long with a heavy dark blue-gray body. Another larva had a body that was green. All had a large "false eyespot" on the top of segment one.

Pupation took place four or more days after the larva was placed in a container with leaves on a bed of peat. The pupa was "yellowish with large lateral eyespots."

Emergence took place 10 to 12 days later. Tveten's notes describe these as "beautiful tersa sphinx" and "elegant moths" with double underlines. He describes the forewing with narrow stripes converging toward the tip. Purplish colors were much prettier than described in the books according to his notes. The hindwing had jagged dots through a yellow background.

An adult was documented in August in Baytown hanging from a planter box during the day.

The following details are taken from Tveten's January 10, 1997, article in the *Houston Chronicle*.

> Leon Hale, in his October 22 column, wrote of watching a creature that looked like "a miniature hummingbird." Hovering at flowers, it displayed a "wide white stripe" across its back. Nowhere in his Peterson's bird book could Hale find such a hummingbird.
>
> Hale contacted local bird expert Bob Behrstock, who suggested that Hale's visitor might be either a coquette hummingbird or a titan sphinx. The former would be unprecedented, since there are apparently no verified records of the tropical coquettes in Texas.

Adult Tersa Sphinx moths.

Variations in Tersa Sphinx larva: brown larva

Green larva

Although the eastern titan sphinx reportedly occurs west-
ward to our state, it, too, would be unusual. In 37 years of
watching Texas lepidoptera, we have never seen this dark moth
with the distinctive white band across the base of its abdomen.

Amazingly, as we were reading Hale's column, we received a
phone call from friends Anne and Fred Speers near Conroe,
who told us of a similar occurrence in their flower garden.
Experienced birders, they also knew a coquette would be
unlikely, but could scarcely believe their visitor was not a
hummingbird.

Yellowish pupa of a Tersa Sphinx moth. See the darker form below.

Other reports trickled in from across the region. Several more "mystery guests" appeared in area gardens, each posing as a hummingbird, each sporting a coquette-like band across its body. At the same time, people from Houston to Corpus Christi were asking us about large green caterpillars with false eyespots

on their bodies that were consuming pentas in their flower beds. Here, too, was a creature we had never seen.

Checking *A Field Guide to the Moths of Eastern North America* by Charles Covell, we discovered titan sphinx larvae feed on plants in the madder family, the Rubiaceae, which includes the pentas. The connection seemed obvious, and we called the

Speers and asked them to check their pentas for caterpillars. Sure enough, there was a huge caterpillar they dubbed "Charlie" methodically consuming leaf after leaf.

A prominent sharp but harmless "horn" at the tip of the abdomen undeniably marked Charlie as a sphinx larva, as did his habit of raising the front part of his body and pulling in his head when disturbed. This characteristic pose was thought to resemble the Egyptian sphinx and gave the family its common name.

The Speers kindly brought us this lovely caterpillar, and we raised it to maturity, certain it would produce the mysterious titan sphinx for us to photograph. After consuming half of our own pentas, Charlie transformed into a bare pupa among fallen leaves, as is the habit of many sphinxes.

We waited impatiently for the moth to emerge and, indeed, we were recently rewarded with a large and lovely sphinx. Embarrassingly, however, it proved to be a tersa sphinx, not the expected titan. Apparently the two species share the same family of larval food plants, and we learned once again it is seldom safe to jump to conclusions where the intricacies of nature are involved.

We were not lucky enough to see the errant hummingbird mimics, but we assume they were titan sphinxes. Instead, we illustrate this column with photos of the larva and adult of the tersa sphinx, solving the riddle of the pentas-loving caterpillars.

Both moths are fascinating additions to Houston's backyard gardens, where they masquerade as darting, hovering hummingbirds.

Range in Texas: Eastern half of Texas with some records in the Panhandle.

When Found in Texas: March to November

Food Plant: Catalpa, penta, hamelia and others.

Tersa Sphinx.

Tersa Sphinx adult and pupa.

Adults frequent nectar-rich plants and fly like a hummingbird. During the larval stage, this species can be thick in Texas. It is literally possible to see hundreds crawling across the road in the right conditions.

Tveten found "hundreds moving across the road in a space of a couple of miles" in Pecos County in May 1981. He wrote that there would be breaks where there were none but then another infestation nearby.

"In the ditches, there were literally thousands of larvae. Standing in one place, I could pick up a jar full. They had absolutely stripped vast stands of a small-leaved gaura, probably accounting for the mass movement. It appeared to be the smaller larvae that were migrating. The big

Larval stage of the White-Lined Sphinx moth.

Adult of the White-Lined Sphinx moth.

ones remained stationary." He added a footnote that read, "I have never seen such an infestation of non-gregarious caterpillars before."

Larvae between Carlsbad Caverns National Park in New Mexico and Pecos Texas were collected and raised. They were found on and fed velvet-leaf gaura, woolly gaura, whitest evening-primrose, and sand verbena.

At Frio Bat Cave in the Hill Country, larvae were found feeding on evening-primrose, gaura, and fiddleleaf nama (*Nama jamaicense*).

Larvae vary depending on the instar. They can be primarily black with bands of yellow when small. A horn is at the end of the abdomen.

Pupation was in leaf litter or sphagnum moss on a surface of dirt two to three weeks after collection.

An adult was found under the lights of a building at Los Fresnos in the Rio Grande Valley. The triangle-shaped adult had a tan line running down the middle of each dark brown forewing. The smaller hindwings were hidden at rest but showed lovely rose in flight. The abdomen was tapered to a sharp point.

Range in Texas: Entire state

When Found in Texas: March to October

Food Plant: Gaura, evening-primrose, sand verbena, fiddleleaf nama

Major Datana (or Azalea Caterpillar Moth) *Datana major* 7905

Adults have tan wings held over the body like a tent. Fine, darker lines cross the wings horizontally. The hairy rounded head is russet. Long antennae are held tightly back against the wings.

Tveten found a mass of larvae on the leaf of a staggerbush in the Kirby State Forest. The larvae were 18 mm long and had black-and-white striped bodies. Their last abdominal segment was red as was the head.

When disturbed, the larvae assumed a U-shaped defensive posture. He noted that the head was raised, tail raised, with most of the body flattened on a leaf.

Mass of larvae feeding on a leaf.

Defensive posture of a Major Datana larva.

Eight days later, the larvae had molted to a black with white checkered pattern. The head, legs, and segment were still red. Scattered large white hairs covered the body.

To pupate, the larvae darkened and then laid on the bottom of their jar. Tveten described them as "short and fat" and that the entire larval skin remains stretched out. "It looks like a dead caterpillar," he wrote.

Range in Texas: Texas Piney Woods east of Dallas and north of Houston

When Found in Texas: August to November

Food Plant: Staggerbush, azalea

Adult form of the Spotted Datana.

Spotted Datana *Datana perspicua* 93-0039 (7908)

Adults have tannish yellow forewings that are decorated with horizontal darker lines. An obvious dot decorates the forewings that are held in a tentlike fashion over the body. Hindwing is light tan to almost yellow. Hairy head is russet.

Larvae have deep red bodies with light stripes and long white hairs. They curl into a U-shape when disturbed.

Defensive posture larvae of the Spotted Datana.

When 32 mm long, the larva had changed to a black body with three yellow stripes on the sides. Tveten noted that there was another yellow stripe just below the spiracles. Each larva was lightly covered in white hairs with a black head.

The larvae ate "voraciously" leading up to forming a pupa. When ready to make the change, they immediately burrowed into loose soil. Skins were left "neatly shed and intact."

Adults emerged in captivity seven to 11 months after pupating.

Range in Texas: Throughout the state except the Rio Grande Valley

When Found in Texas: April to October

Food Plant: Flame-leaf sumac

Adult for of the Angulose Prominent:

Angulose Prominent *Peridea angulosa* **7920**

"Perfect bark camouflage" was how Tveten described this moth. Gray and white mottling would allow this moth to disappear on nearly any oak tree.

Eggs are large and white with a high dome, like half a ping-pong ball. Tveten found no surface sculpture at 6x magnification. Eggs are laid singly versus in masses.

Larvae emerged nine days later and started feeding on oak leaves. In captivity, Tveten found that the larvae ate only willow oak and would consume one leaf a day. Larger

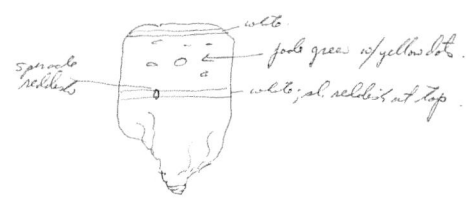

larvae ate four to six oak leaves a day. As the larvae grew, they lived along the midrib of each leaf and fed on the leaf blade.

Tveten described a larva as "jade green with whitish stripes and yellow dots. Head green; large and sloping." Another specimen was "green with double pale-yellow dorsal stripes. Bright yellow lateral stripe." At 26 days old, the larvae were 32 mm long. Two days later, they were 40 mm long and 5 mm wide.

Pupas were large and covered in a few strands of silk.

Range in Texas: Dallas into the Hill Country and east to the Houston area

When Found in Texas: March to October

Food Plant: Willow oak and other oaks

Adult White Furcula moth.

White Furcula *Furcula borealis* **7936**

A furcula is a member of the prominent moth family with a rounded "helmet" of hairs on the head. Wings are held "tentlike" over the body. The white forewing has dark spots and a brownish middle that resembles a saddle.

Larvae have an unusual shape that could be startling. Two tentacles extend from the abdomen forming "whip-tails." The body changes from

Larva showing defensive posture.

green to yellow to rose with a brown area along the back called a saddle. Larvae can measure 27 to 40 mm, or 1 to 1½ inches.

Larvae were found on top of black cherry leaves. They were "eating voraciously."

Tveten noted that in "prepupation 'whips' have swelled." Cocoons varied. One spun an oval cocoon on a leaf. One spun a cocoon that blended perfectly with a stick. "One studded cocoon surface with small lichens and lichen-covered twig."

Moths emerged 10 days later.

Range in Texas: Northeast Texas, Dallas, Austin, and San Antonio. Tveten found these in the Little Thicket in San Jacinto County.

When Found in Texas: April to August

Food Plant: Black cherry

Adult form of the Drab Prominent.

Drab Prominent *Misogada unicolor* 7974

This species is appropriately named. Drab, brown wings held like a tent over the body. A freshly emerged adult might have a greenish sheen on the wings.

Larvae are bright green with prolegs elongated into double "tails." There is a white stripe on the top of the body decorated with small red spots. Faint, thin yellow stripes run along the side above the spiracles. The spiracles can be ringed with yellow. A large larva measured 38 mm long and 6 mm wide, or 1½ inches long by ⅕ inches wide.

When disturbed, the larvae coil up on the underside of a leaf near the petiole. "It looks as if they would fall off, but they cling tightly and show no tendency to drop," wrote Tveten.

Pupating larvae turned deep rose-pink in color. The "tails" were turned under. The cocoon was flat and flimsy on a leaf. Others were observed to spin on a folded leaf or sew together corners of a leaf.

Larva of the Drab
Prominent with
recently shed skin
nearby.

Instar of the
Drab Prominent
larva showing
brown color.

Instar of the Drab Prominent with green color.

Adults emerged six to nine months after pupating.

Range in Texas: East Texas to the Hill Country

When Found in Texas: March to October

Food Plant: American plane-tree or sycamore

Adult of the Variable Oakleaf Caterpillar Moth.

Variable Oakleaf Caterpillar Moth *Lochmaeus manteo* **7998**

The species is a fairly large gray prominent moth. The forewing has a dark triangle on the costa. A pale spot on the forewing has a dark center. The remainder of the wing is banded but this can be worn and obscured.

Tveten found several small larvae in moss on a post oak tree in his Baytown yard. The 7 mm long larvae were yellow green with red markings. Each had a mottled head covered with hairs. Their bodies had hairs, but mostly on the dorsal side.

At 20 mm long, the larvae were "really beautiful with red and yellow markings." At maturity, the larvae had solid red backs, and some had greenish backs.

Range in Texas: Dallas to San Antonio and east

When Found in Texas: April to October

Food Plant: Post oak and other trees

Different markings and the larvae matured and developed.

Mature larva of the Variable Oakleaf Caterpillar Moth showing red back.

Double-Lined Prominent *Lochmaeus bilineata* **7999**

This species has the typical profile of a prominent moth with a raised area of hair over the head and forewings held over the body like a tent. Two double lines cross the forewings. Tveten noted that Gulf Coast specimens of this species are paler and smaller per his reference book from Charles Covell.

Tveten observed that a larva in captivity laid quietly in the bottom of a jar and formed a bare, shiny, dark brown pupa.

An adult emerged one month later.

Range in Texas: East Texas from Dallas to Hill Country to Rio Grande Valley

When Found in Texas: February to December

Food Plant: White oak and other deciduous trees

Adult Double-Lined Prominent showing raised hairs over the head.

Stages of larval development in the Double-Lined Prominent.

Red-brown saddle marking on
the Double-Lined Prominent.

Adult Unicorn Prominent moth.

Unicorn Prominent   *Coelodasys unicornis*   8007

This species is also known as Unicorn Caterpillar Moth by Tveten or Variegated Prominent. Adults are deep brown with wings held tight against the body often looking like a broken stick. There is a peak over the head which is really a group of hairs on the trailing edge of the forewings. Cryptic coloring, tight wings, and the peak on the head allows this moth to mimic a twig or broken branch.

Tveten found larvae in various places in the East Texas Piney Woods from June to October. He observed that, "The camouflage is amazingly good. The larva chews out part of the leaf and sits in the opening."

Larvae were noted to be various sizes but were identical in color and pattern. Overall earthy brown colored, the larva has a spike on the fourth segment. Tveten wrote, "These are brown with second and third thoracic segments green (just ahead of 'lungs') with white chevrons on back just ahead of anal 'lump.'"

Small cocoons were woven between two leaves on a twig. Tveten also observed cocoons woven under leaves in the bottom of the container. Cocoons formed two to 10 days after being moved into captivity.

Larva chewing leaf and sitting in the opening.

Adult Unicorn Prominent from above.

In one instance, Tveten noted that the "larva remained unchanged in cocoon for many weeks." This specimen then pupated and emerged fairly quickly.

Adults emerged 10 days to six weeks after forming pupas.

Range in Texas: Most of Texas except the Panhandle and far western areas

When Found in Texas: January to November

Food Plants: Black cherry, hawthorn, flowering dogwood, American hornbeam, Carolina cherry-laurel, Carolina buckthorn, blackberry, buttonbush

Adult Scarlet-Winged Lichen Moth.

Scarlet-Winged Lichen Moth *Hypoprepia miniata* **8089**

What can be more beautiful than a dark gray moth with bold scarlet lines? Forewings are broad and held overlapping the body. The ½ inch long forewings conceal hindwings that are gray with scarlet toward the body. Forelegs are striped scarlet and black.

Tveten found this moth at a blacklight in Concan, Texas. An adult was placed in a jar and it laid 60–70 eggs. Tveten described these as "like tiny black pearls." Each was round shiny, and lightly dimpled like a golf ball.

Unfortunately, all the larvae dried up as they changed from the first instar to the second.

Range in Texas: Dallas area through the Hill Country. Scattered reports east and west of this area.

When Found in Texas: April to October

Food Plant: Lichens

Larva of Scarlet-Winged Lichen Moth on lichen.

Ornate Moth (now called Ornate Bella Moth)
Utetheisa ornatrix **8105**

Adults have a long, narrow profile. Forewings are peachy white and bordered with peach and black spots. Long antennae are black and legs are black. Eyes are large and black.

Tveten found two larvae feeding on rattlepod (Crotolaria), which is in the legume family, at Falcon Lake in Starr County, Texas. Larvae were burnt orange with wide black bands on each segment.

In captivity, the larvae pupated in five days. Cocoons were "flowing silk" suspended from the jar lid.

Adults emerged nine days later and were described as "lovely pinkish moths."

Adult Ornate Moth.

Larvae of Ornate Moth feeding on rattlepod.

Range in Texas: Scattered reports from the eastern half of the state from Dallas/Fort Worth through Austin and San Antonio to the Lower Rio Grande Valley. More common in Florida and Mexico.

When Found in Texas: May to December

Food plant: Rattlepod or rattlebox. Others report feeding on clover, elm, cherry.

Salt Marsh Moth (a.k.a. Acrea Moth) *Estigmene acrea* **8131**

The Salt Marsh Moth is a handsome creature with white wings lightly speckled with black. The head is furry, and legs are banded in black and white. The adult's abdomen is orange and black.

Larvae have long hairs at all stages. The larva's body color changes from pale to yellow stripes on a dark base.

Tveten encountered this species in the Houston area, central coast, Rio Grande Valley, and west to Monahans Sandhills State Park.

While raising larvae, Tveten found that this species is cannibalistic: "These are cannibalistic on the pupae of others. Two larvae destroyed a single pupa. Then one spun a rather good cocoon and pupated. The other ate a large hole in the cocoon and started to attack the pupa." He saved

Adult of the Salt Marsh Moth.

Long hairs and pale color on early stage larva.

that pupa from being eaten but it emerged later with a piece of wing missing possibly from the earlier attack.

Tveten described small salt marsh larvae as looking "something like fall webworms." He noted that the larvae were often found in groups with different stages of development. On one November outing at Armand Bayou Nature Center near Houston, Tveten noted that these were the most abundant caterpillars seen.

Two to three weeks of development yield a larva that can be identified as a Salt Marsh Moth. At this point the hairy larvae have a black base with yellow lengthwise stripes down the body.

The larval stage was observed to be 28 to 37 days. Larvae spun cocoons and emerged 14 to 16 days later.

Tveten was watching an adult female Salt Marsh Moth resting on bluestem at Goose Island State Park on the central coast: "We watched as a large, colorful, jumping spider pounced. It buried its fangs just behind the head, with orange fluid seeping out of the moth." Both critters were captured and placed in separate jars. Both survived. The female eventually laid eggs in the jar.

Color change noted in the larva of a Salt Marsh Moth as it aged.

Range in Texas: Most of the state except the arid deserts

When Found in Texas: January to November

Food Plant: Clover, blackberry, rattlebean, Chinese tallow, common persimmon, mulberry, whitestem wild indigo, spectacle pod, yellow mustard, yucca blooms, eupatorium

Adult Virginian
Tiger Moth.

Virginian Tiger Moth *Spilosoma virginica* **8137**

The adult Virginian Tiger Moth is hard to ignore. Its solid white wings have one tiny black speck on the upper surface of the hindwing and one to two smaller specks on the upper forewing. The plump body is covered with white hairs.

Tveten found the larvae of this species in Texas from Wallisville, Baytown, Houston, and south to Goliad and Three Rivers. The larvae were found in wilderness as well as in urban habitats.

These "inchworm" larvae ranged from 2.5 to 45 mm (0.1 to 1.7 inches). All were green with colors ranging from light green to bright gray-green to pale greenish white. Each was covered with "very frilly" white hairs and had a few dark hairs mixed in. The heads were peach colored to light yellow-orange.

Larvae were found on water hyacinth, climbing vines, cassia, and ash in the wilderness, and Virginia sweetspire (*Itea virginica*) in an urban setting. They skeletonized the leaves of these plants as they fed.

In one instance, the larvae cut pieces of leaves and fixed these to their backs. Tveten described it as "super camouflage."

Early instar of the Virginian Tiger Moth with white hairs.

Larvae of Virginian Tiger Moth skeletonizing leaves as they feed.

Later instar of the Virginian Tiger Moth showing darker color.

Adult with wings raised to show body detail.

Tveten noticed that in captivity the larvae became darker with each instar. The long white hairs changed to dark hairs, the body became darker brown or reddish, and the head darkened to orange-brown. One group of larvae remained white, though, until they pupated.

Cocoons were spun with the larva's body hairs and were described as "loose" by Tveten. He noted that the shape was a small oval and some were devoid of hairs.

Adult moths emerged eight to 12 days after pupating. One group of larvae pupated in October and did not emerge until late April or six months later.

Range in Texas: Eastern half of the state from Dallas/Fort Worth through Austin and San Antonio to the Lower Rio Grande Valley. Scattered reports from the rest of the state.

When Found in Texas: March to October

Food plant: Water hyacinth, cassia, ash, Virginia sweetspire

Fall Webworm *Hyphantria cunea* **8140**

What homeowner has not dreaded the sight of webby nests covering an entire tree in the yard? Infestations are not limited to autumn in Texas.

Tveten observed in October 1989, "Webworms absolutely decimating sweetgums (and some others) along highway and at Little Thicket. Entire trees webbed and denuded. I've never seen them so bad." Yet, other times he noted a single new web or maybe old webs hanging in trees.

Eggs were tiny green spheres laid in a mass. Females were observed to lay 50–60 eggs at a time in a large mass. Eggs were laid in a pattern of tightly packed rows and covered with white hairs.

Larvae emerged five days after eggs were laid. First instar was pale green with a dark head and sparse white hairs. Within a month, they had a pair of dorsolateral black spots on each segment with several bands of white hairs.

Larvae "twitch front part of body violently when disturbed." The mass will skeletonize a leaf. Toward the fourth and fifth instar there is less time spent on building the web and more time spent on eating leaves.

Pupation is in a flimsy cocoon on the ground or buried in grass. Adults emerge a month later.

Mass of Fall Webworm larvae feeding on leaves.

Females lay many eggs in a tight mass covered in white hairs.

Early stage instar of Fall Webworm larva.

Larvae darken as they age.

"Males tend to emerge first," Tveten noted.

Adults are white with varying amounts of spotting on the wings. Some adults are completely white with no spotting. Once, when an adult emerged with immaculate white wings, Tveten wrote, "I don't see how I could have gotten another species in the system." He later learned that adult fall webworms in the south have more spotting than those in the northern United States. Tveten later wondered if overwintering broods have more spots after he found two adult females at his porchlight in March. Both were heavily spotted with black and dark gray.

Adults can be confused with similar looking tiger and leopard moths. Fall webworm adults have dark bands on the base of their forelegs.

Tveten separated a large group of larvae that he found on mulberry into three jars. The group placed in a jar with pecan leaves converted to those leaves readily. A group placed in a jar with red mulberry leaves converted to those leaves "moderately easy." A third group placed in a jar with white mulberry leaves ate but not as well.

Range in Texas: Eastern half of the state with some from Lubbock and Fort Davis

When Found in Texas: All year

Food Plant: Flowering dogwood, pecan, sweetgum, mulberry, American plane-tree, sycamore

The spotting on adult Fall Webworm moths can be variable.

Detail of adult moth.

(no common name) *Macalla glastianalis* **5576.1**

There are few records of this moth in Texas. In the past, it had the Latin name *Incertae sedis*.

This small moth with deltoid wings is a member of the Pyralid Snout moth family. Adults have bicolored brown and silver wings. A black line separates the two colors. There is a rounded "crest" on the head.

Tveten found several larvae in a stand of Coyotillo in Alamo, Texas. They were building nests by webbing together several terminal leaves. Each nest had one larva.

Larvae were yellowish with dark stripes and less than 1 inch long.

He collected several and two pupated in the bottom of their jar. One pupated in webbed leaves.

Adults emerged 24 days after Tveten first found them.

Range in Texas: Extreme Rio Grande Valley to Corpus Christi area

When Found in Texas: Year round

Food Plant: Coyotillo

Larva of the *Macalla glastianalis.*

Giant Leopard Moth *Hypercompe scribonia* 8146

Giant Leopard Moths are easy to identify in the adult stage. The rounded triangular-shaped wings have a white base with black circles. The interior of each circle is white. The hindwings are usually tucked but are white with black markings. Legs are white with black knees. The thorax is plump with orange and black on the top surface.

Tveten found larvae and adults at his home in Baytown, Texas, February to December. He also found larvae in Wallisville and the Little Thicket.

Larvae have "black, spiny segments with orange between." He described one larva as having "broad bands of black and rusty orange with black spines." Children would call this a "woolly worm."

Larvae were found on a variety of plants including American beautyberry, asparagus fern, and yaupon. In captivity, larvae feed well on willow, peppervine, dandelion, sow thistle, cooper daisy, air potato, Boston fern, asparagus fern, and plantain. In captivity, they did not eat American beautyberry, baccharis, salvia, violet, or oak when offered.

One larva was found under the bark of a dead tree in March. Tveten speculated that it had overwintered in that location. Mature larvae were found in February at his home and Tveten again wrote that they must have overwintered as larvae.

Adult of the Giant Leopard Moth.

Giant Leopard Moth larva showing orange between each segment.

Larvae of the Giant Leopard Moth in various stages of development feeding on fern.

Larvae in captivity pupated in about a month but several stayed in the larval stage for two to three months. Larvae found in February pupated rapidly and emerged in 13 days. Cocoons were formed with leaves bound with webbing.

One adult found in late March began laying eggs within four hours of being captured and placed in a jar. The eggs were not fertile though, and never produced young.

Range in Texas: Common in the eastern half of the state from Dallas/Fort Worth through Austin and San Antonio to the Lower Rio Grande Valley. Scattered reports from Midland/Odessa area.

When Found in Texas: All year but primarily in the warmer months

Food plant: Willow, peppervine, dandelion, sow thistle, cooper daisy, air potato, Boston fern, asparagus fern, and plantain

Large Leopard Moth or Mexican Leopard Moth
Hypercompe muzina   (**no Hodges number**)

(This species has been added for the historical record. It is not supposed to be north of Mexico.)

In January of 2000, Tveten found a larva in Cameron County at the Inn at Chachalaca Bend. The larva pupated "almost immediately" when brought inside and placed in a jar. The pupa was very dark and in the base of the jar.

The larva looked like a Giant Leopard Moth but was smaller and less heavily spined. Another was located in March of 2000 and was described as "spiny black larva with red joints."

One adult emerged and was simply described as "much like a Giant Leopard Moth."

These descriptions and photos are offered as historical records. Tveten was very familiar with the Giant Leopard Moth but felt these specimens were something different.

Larva described in Tveten's notes.

Adult specimen that Tveten mentioned in his notes.

Range in Texas: One record at Estero Llano Grande State Park in the Rio Grande Valley

When Found in Texas: January to March

Food plant: Unknown

Variations in the adult form of the Nevada Tiger Moth.

Nevada Tiger Moth *Apantesis nevadensis* (was once *Grammia nevadensis*) **8179**

This is a handsome moth with striking markings. Adults have a black, furry head with black antennae and legs. Wings are cream colored with bold, black markings. A series of thin white lines form a "W" pattern on the submarginal forewing that is only visible when all wings are spread. Hindwings are salmon pink with black dots in the submarginal area. Each wing is fringed in creamy white.

Larva of the Nevada Tiger Moth found by Tveten.

Tveten found Nevada Tiger Moth larvae in Fort Lancaster State Park in Crockett County, Texas in May 1981. Fort Lancaster is off Interstate 10 between Ozona and Fort Stockton and currently thought to be the moth's historic range.

"Many crawling around on the ground," Tveten wrote. "They move very rapidly, much more so than any caterpillar I have previously encountered."

Larvae fed on senecio, coneflower, and small-leaved gaura for 10 days before pupating.

The pupae were bare.

Tveten made a note that his photos did not look like those in Cornell or Holland, the standard reference at the time.

Range in Texas: Historic records from West Texas from the Panhandle to near Laredo. iNaturalist indicates no current records.

When Found in Texas: May

Food Plant: Senecio-like ragworts and groundsels, coneflower, and small-leaved gaura

Adult Davis' Tussock Moth.

Davis' Tussock Moth *Halysidota davisii* 8205

This moth is beige with very few markings. Wings are held back against the body and overlap.

Tveten found approximately 40 larvae infesting a catclaw acacia in Limpia Canyon.

Larvae are dark with a thick coating of white hairs. Red tufts are prominent on both ends.

Larvae placed in captivity in October emerged six months later in June.

Range in Texas: El Paso to the Big Bend area plus the Panhandle

When Found in Texas: June to October

Foot Plant: Catclaw acacia

Larva of the Davis' Tussock Moth.

Schaus' Tussock Moth *Halysidota schausi* **8205.1**

The adult of this species is strikingly elegant. Pale, nearly translucent forewings are pale peach with delicate grayish markings. Forewings are held close to the body concealing a smaller whitish hindwing. The body is thick with hairy grayish head and thorax and rusty colored abdomen.

Tveten found hundreds of these larvae stripping hackberry leaves around Lake Corpus Christi State Park one May. He wrote that, "when alarmed (they) twitch and flare white tufts at head. Also flash orange section between thoracic segments."

Eggs were laid in a tightly packed groups of 60–70 eggs. Each was round and very white. Eggs hatched in eight days.

The first instar was a "fairly large, pale green caterpillar." In a few days, they molted from hairless to hairy. Eight days later, at the third instar they took on the look of tussock caterpillars. Seventeen days later, the larvae were mature with "red-orange around base of white tufts."

Tveten noted a high variability in larvae colors from gray, brown, white, and yellow. All had several long tufts of hair from the head and two long tufts from the base of the abdomen.

Adult Schaus' Tussock Moth.

Early stage of larvae as they feed on a leaf.

More mature larva of a Schaus' Tussock Moth.

Variability in larvae of the Schaus' Tussock Moth.

Cocoons were spun of hairs with a large brown pupa. Emergence was five months later in captivity.

Range in Texas: Houston to San Antonio and down to the Rio Grande Valley

When Found in Texas: March to December

Foodplant: Netleaf hackberry, spring hackberry, sugar hackberry

Milkweed Tussock Moth *Euchaetes egle* **8238**

A robust pale tan moth with forewings that are ¾ inch long. Hindwing is the same color and both are slightly mottled. The thick abdomen of the moth is orange with a row of tiny black dots on the dorsum.

Larvae are black with white tufts near the end both dorsally and to the sides. Older larvae have six orange tufts on the dorsum.

Tveten observed small larvae in the first or second instar feeding close together on a Mexican milkweed plant in a garden. The larvae "absolutely destroyed" the plant, eating leaves, flowers, and seed pods. "They ate the outer shell of its pods and all the green seeds, leaving only the silky fluff," he wrote. As the larvae grew, they spaced out to several other milkweed plants in the garden.

Maturing took two weeks in one instance. Cocoons were small black ovals of hairs.

Adults emerged 10 days after spinning cocoons in the summer.

Range in Texas: All the state except the Panhandle

When Found in Texas: March to September

Food Plant: Milkweed

Adult Milkweed Tussock Moth.

Older larva of a Milkweed Tussock Moth.

Yellow-Collared Scape Moth   *Cisseps fulvicollis*   **8267**

Two of these moths showed up under Tveten's porch light in Baytown, Texas. He described them as having, "black wings, blue bodies, and yellow collars." Forewings are ½ inch long and held elongated against the body.

This is a diurnal moth and might be found during the day feeding on flowers.

The moths laid eggs a day after being placed in a jar. The eggs hatched four days later.

Larvae fed on coarse weed grass from the yard. They fed sparingly on St. Augustine. Eighteen days after hatching, they were approximately 1 inch long with clusters of spiny hairs.

A month after hatching, the larvae spun very flimsy little cocoons in grass. These cocoons were mostly made up of larva hairs. They pupated a day after spinning. The pupa was yellowish with dark markings.

Two weeks after spinning, adults emerged. Tveten noted that this group of progeny had orange collars where their parents had yellow collars.

Range in Texas: Entire state except arid deserts of the west

When Found in Texas: All year

Food Plant: Coarse grass with lichens and spike-rushes (in the literature)

Adult Yellow-Collared Scape Moth.

Larva feeding on blade of grass.

Adult Southern Tussock Moth camouflages on tree bark.

Southern Tussock Moth   *Dasychira meridionalis*   8298

This moth can be perfectly camouflaged on tree bark. Mottled wings are held in a delta shape. Gray, brown, and black are the main wing colors but there is a whitish patch in the costa of the forewing. Two long, hairy legs are held forward near the face. A smaller set of legs is held away from the body near the top of the forewings.

Tveten found this species regularly in his Baytown yard. Adults were found around the porch light. Larvae were found crawling on the house, on the trunk of a willow oak, and on leaves of the oak. He wrote, "This is a common species I see every spring."

Eggs were pale aqua or turquoise colored. The top of each was flattened with a darker glossy central area. Each was covered with "thin, clear film; as if liquid spread and is beginning to harden." Hairs from the moth were lightly sprinkled on each egg.

Eggs were laid in tightly packed masses on a twig. A female in captivity laid several masses rather than one large mass of eggs.

Hatching took place 11 days later. Larvae were brown with scattered hairs. Over time they took on the classic "tufts" of a tussock moth. The "tufts" were

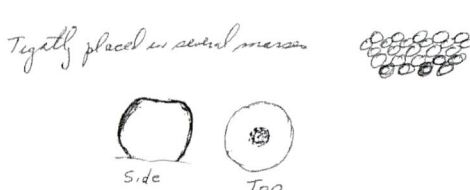

Tightly placed in several masses

Side

Top

Larva form of the Southern Tussock Moth.

clusters of hairs protruding from the side of the larvae. Two long, black tufts near the head resembled antennae and two protruded from the tail. Thicker clumps of hair developed on the thorax.

The larval stage can last a month or two with transitions through five instars. Mature larvae can be 30 mm long including the tufts.

Cocoons were loose and covered in gray hairs. Cocoons were found on oak leaves or in the bottom of the jar in captivity.

Moths emerged 10 to12 days after pupating. In one instance, a moth emerged seven months after forming a cocoon.

Tveten called one specimen Kerrville Tussock Moth or *Disychira meridionalis kervillei*. Found in Real County, this was a "very large Tussock female with black lines on wings." In the group that Tveten raised, one female had a black line on the forewing and one did not.

Range in Texas: Dallas, Austin, San Antonio, Corpus Christi, and east

When Found in Texas: March to October

Food Plant: Willow oak, post oak, burr oak, and white oak

Variations in adult Southern Tussock Moth wing patterns.

Kerrville Tussock Moth that Tveten raised with a "black line on the forewing."

Kerrville Tussock Moth without the black line.

Larva collected in Real County, Texas.

Faint-Spotted Palthis *Palthis asopialis* **8398**

This "stealth fighter" shaped moth has long, upturned palps. Forewings are ½ inch long, mottled brown, with a lighter postdiscal area on both wings.

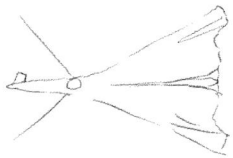

Tveten wrote that this moth was "shaped like a Concorde airplane." Wings are brownish with a wrinkled appearance or supple folded wings. In another instance he wrote that the forewings were pleated.

Eggs were laid singly in a jar.

Larvae fed for a bit on new leaves of a willow oak. They ceased eating on oak but fed slowly on dewberry leaves, which they skeletonized. As the larvae grew they continued to feed on dewberry but only the old, dead, and withered leaves. Tveten noted that they ignored the newer, more tender foliage.

Larvae were olive brown with irregular segments. Another set of larvae were dark brown with slightly lighter lateral broken stripes. The anal plate on the last segment formed a pronounced bump.

Nearly a month after hatching, one of the larvae formed a bare pupa in an old, withered, dried leaf.

Pupae were formed in loose silk. Adults emerged seven days after pupating.

Range in Texas: Eastern half of Texas

When Found in Texas: February to December

Foodplants: Dewberry, willow oak, beans, corns, oats

Two different adult Faint-Spotted Palthis moths.

Green Cloverworm Moth   *Hypena scabra*   **8465**

This medium-sized dark, deltoid moth forms nearly a perfect triangle at rest. Wings vary from light gray to darker mottled brown with a bit of a pattern. Eyes are large and bulging. Antennae are long and threadlike.

Tveten noted that this species of moth was common on early December nights under his porch light in Baytown. He raised many but also fed them to his tarantula.

Eggs were pale green flattened spheres with shallow ribbing. Hatching took place about six days after being laid.

Larvae were bright green when ¼ inch long. At 18 mm, or nearly ¾ inch long, the slender, bright pale green larvae had developed light

Larva of the Green Cloverworm Moth.

Adult variations in Green Cloverworm Moths.

lines. Larvae in captivity fed slowly and skeletonized the leaves of dewberry. Larvae raised in captivity on senna eventually died.

Pupae were surrounded by a few strands of silk in leaves. Adults emerged about 20 days later.

Range in Texas: Entire state.

When Found in Texas: All year

Food Plant: Dewberry, bear clover, black medic, clover, alfalfa

Adult Moonseed Moth show "lumps" in profile.

Moonseed Moth *Plusiodonta compressipalpis* **8534**

The profile of this moth appears "lumpy" per Tveten. The ½-inch-long forewings have a purplish sheen with golden markings. The purplish hue can also be rosy in color.

Eggs were laid in patches on leaves a day after the adult was placed in a jar. The pale greenish, round eggs hatched five days later and the larvae

Fourth instar larva of a Moonseed Moth resembling bird droppings.

immediately started feeding on moonseed vine. The leaves of the vine were skeletonized.

Larvae in the first instar are tiny, pale green. At the second instar, they are green and obviously "looper" caterpillars. During the third instar, they are emerald green with a single black spot on each side of the abdominal segments and dorsolateral. By the fourth instar, larvae are mottled brown and white looking like bird droppings. At this stage, Tveten noted that they resembled young swallowtail butterfly larvae.

Spinning began 13 days after hatching. Tveten noted that they first make a wall on each side of the body and then make a spindle-shaped cocoon on a stem. Cocoons were also attached to a stick and looked like elongated bumps. The cocoon incorporated wood fibers to aid in camouflage. Pupas were long, waxy, and dark.

Adults emerged 13 days later through a slit in the cocoon.

Range in Texas: Eastern two-thirds of the state

When Found in Texas: February to December

Food Plant: Snailseed vine, lime prickly ash, spring hackberry

Hieroglyphic Moth *Diphthera festiva* **8560**

This moth looks like a piece of sculpture versus a living creature. Wings are held like a tent over the body. Both are light orange with dramatic black lines. The forewing submarginal band is edged in three rows of tiny black dots.

Tveten found his first specimen of this species on a gas pump in Baytown. He captured and photographed it, but noted it was a "wonderful, lovely moth; although not as bright orange-yellow as Cornell specimen."

Years later he found numerous large larvae stripping two different bushes at Laguna Atascocita in Cameron County. The larvae were 45 mm, or 1.7 inches long and "fairly heavy." The larvae were black and white with some barring on the dorsum. They had a red head, black crest stripe, and were orange below the spiracles.

Captured larvae were fed melochia or tea bush.

Adults emerged in July after eight months in the cocoon.

Range in Texas: Coast of Texas inland to Austin, San Antonio, and Laredo

When Found in Texas: April to December

Food Plant: *Melochia tomentosa* (also called pyramid bush, tea bush)

Adult Hieroglyphic Moth.

Larva of the Hieroglyphic Moth.

Black Witch *Ascalapha odorata* **8649**

This moth is often mistaken for a bat. Its dark wings can span 4–6 inches and are held flat a wall or tree at rest.

Adult males appear to have black wings but these are really intricately marked with lovely patterns and shapes. Females are a bit lighter with a prominent white band through the middle of each wing.

Range in Texas: Throughout the state

When Found in Texas: All year in southern Texas

Food Plant: Cassia and catclaw

Adult Black Witch moths are often mistaken for a bat.

Lunate Zale Moth *Zale lunata* **8689**

A Lunate Zale Moth could easily blend in with the craggy bark of an elm tree, sweetgum, or oak. Its wings are held flat away from the body with all four wings visible. Each is mottled grayish brown. The robust body has a thickening at the base of the head.

Tveten encountered this species in various stages of its development over 30 years. It was a regular around his home in Baytown, Texas, giving him a chance to become familiar.

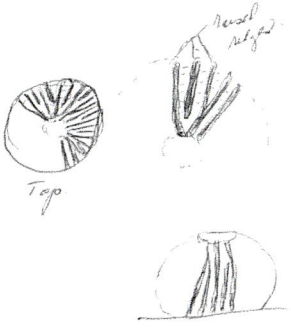

One of those moths laid approximately 40 eggs within a day of being placed in a jar. Each "almost round" jade green egg had about 37 to 40 ridges of "varying length around (the) circumference." Some of those ridges were banded.

Eggs hatched in four to five days with a slender looper caterpillar emerging. The tiny larvae were slender, pale tan, with two pairs of prolegs and anal prolegs. Tveten noted that these "loop very rapidly around jar—hanging by threads of silk."

Variations in the wing patterns of Lunate Zale Moths.

Two Lunate Zale Moth larvae on a twig showing the projection on the anal end.

Maturing larvae grew rapidly to 35 mm long in 20 days. Larvae in this later instar were dark brown with a paler dorsal stripe and longitudinal stripes in shades of brown. Tveten noted dark spots or marks on the ventral side. A projection was noted on the eighth abdominal segment or anal end. The larvae rested with prolegs and true legs displayed out to the side.

Elm or cherry laurel leaves were pulled together and "stitched along the sides" to house the pupa stage. The cocoon was held between the leaves that were "tightly stitched together."

Adults emerged seven to 19 days later.

Range in Texas: Primarily in East Texas from the Louisiana border east to Dallas/Fort Worth and San Antonio. Scattered in the rest of the state.

When Found in Texas: Year round

Food Plant: Cedar elm, willow oak, cherry laurel, American sweetgum

Tveten used the words "large" or "fairly large" to describe this moth under his porch light in Baytown. This mottled brown moth has ¾-inch-long forewings. Two black spots on the inner margins of the forewings resemble eyes. Wings are held to form a triangle with the head forming a definitive point. A submarginal darker band crosses both wings.

Eggs were laid singly in groups of 20 to 60 in a jar when placed in captivity. The eggs were tiny grayish spheres with light sculpturing. They hatched in four days.

Larvae were loopers. At 12 days, larvae were 35 mm long and slender. They were tan with various shades of longitudinal stripes. Each had two pairs of prolegs and anal claspers. A month after hatching, the larvae were 55 mm long and 6 mm wide. They were still tan with long stripes. "Black bars on each side of dorsal between abdominal segments 1–2 and 2–3. Only show when larva is 'looped,'" per Tveten's notes.

Larvae from the same female grew at different rates.

Two months after hatching, cocoons were made by webbing grass stalks together.

Tveten documented this species five times over 20 years. He noted that it looked like a Texas Mocis (*Mocis texana*) but always referred back to Charles Covell's book (page 170) to confirm his identification.

Range in Texas: East Texas west to Dallas and Uvalde south to Rio Grande Valley

When Found in Texas: Year round

Food Plants: St. Augustine grass and other grasses

Variations in wing pattern on Withered Mocis moth.

Black bars on the abdominal segments of a mature larva that Tveten noted.

Larva of The Betrothed Underwing moth.

The Betrothed Underwing   *Catocala innubens*   8770

This underwing moth has the typical mottled forewings of underwings. Each forewing is colored with dark on the outer edge and lighter on the inner edge. Colors can be variable from dark brown to tan, or dark grayish to light grayish. All color variations have a lighter patch on the apex of the forewing and a somewhat lighter outer margin.

Tveten was given eggs to raise by a friend from Iowa in the winter of 1996. He was told the food plant was honey locust. The eggs were refrigerated until late March 1997.

The eggs sat on a damp paper towel and "hatched late" according to Tveten's notes.

Larvae grew but very slowly on honey locust. They eventually became "very pretty" and "highly colored," reaching lengths of 42 mm long.

Only one eventually pupated but Tveten did not record if it emerged as an adult.

Tveten did question why the larvae died "in spite of getting new locust leaves. Is there something in locust that prohibits cutting and storing leaves?"

Range in Texas: Dallas area with scattered reports from the Austin and Houston area

When Found in Texas: April to July

Food Plant: Honey locust and black walnut

Ilia Underwing *Catocala ilia* **8801**

The mottled forewings of the Ilia Underwing can be variable. The grayish brown wings are darker near the body and lighter in the middle of the wing. A zigzag pattern separates the submarginal area of the wing. A white spot on the wing can be nearly absent to prominent.

As in all underwings, the underwing of the Ilia is beautiful with red and black circular bands. The outer margin of the hindwing is edged in white.

Tveten found an Ilia under his Baytown porch light in July of 1991. He described the adult as a "very large, beautiful red-and-black underwing."

He was given a larva by a hiker while on the Boot Springs Trail in Big Bend National Park in April 1992. The larva was crawling on the trail and measured approximately 2 inches. Tveten described it as, "Olive green with brown mottled pattern. Wide, flattened. Appears ready to pupate."

Five days later, May 4, 1992, the larva had pupated in a jar. The pupa was bare, with a heavy white coating.

The adult emerged in late summer but Tveten was traveling so we do not have an exact date.

Tveten was given eggs to raise by a friend from Iowa during the winter of 1996. The eggs were refrigerated and allowed to warm the following February. The eggs hatched within a day. The tiny larvae were "tiny, threadlike" and resembled inchworms. They "gallop wildly around jar."

Post oak was a successful host plant for the larvae as they grew. They were 40 mm long and feeding "voraciously" 14 days after hatching. A month after hatching, they were 45 mm long. Larvae changed from a camouflage, spotted pattern to "very heavy, spiny."

Pupation began five weeks after hatching. The two surviving larvae spun in leaves tied with silk.

Emergence was three weeks later. The total time from egg to adult was two months.

Range in Texas: East Texas from Dallas, through Austin and San Antonio, down to near Del Rio

When Found in Texas: April to July

Food Plant: Post oak; did not like willow oak

Wing variations in Ilia Underwing moths.

Larva of the Ilia Underwing.

Larva of the Ilia Underwing moth.

Girlfriend Underwing moth camouflaged on wood.

Girlfriend Underwing *Catocala amica* **8878**

This moth has gray mottled forewings with a faint zigzag pattern at the submarginal area. A darker scalloped pattern separates the basal and discal part of the forewing. The underwing is primarily orange with a sharply delineated patch in the postdiscal area.

In April 1991, Tveten had a girlfriend underwing fly into his garage through the open door. He was excited in his notes: "Moth flying in garage last night! Small underwing with black border on rear red-orange wing."

He encountered this species again in April 2001 while staying in Leaky along the Frio River in Real County, Texas. A larva was found on his cabin wall. Before he could get photos, though, the larva spun a cocoon.

An adult emerged 25 days later. Tveten noted that it was a small moth with "red hindwing with only outer black band."

Range in Texas: East Texas to Austin and the Rio Grande Valley

When Found in Texas: April to June

Food Plant: Oak

Girlfriend Underwing moth showing colorful underwings.

Larva of the Girlfriend Underwing moth.

Adult Cabbage
Looper Moth.

Cabbage Looper Moth *Trichoplusia ni* **8887**

"Very dark gray-brown moth" was how Tveten described an adult cabbage looper. The inch-long, mottled forewings are marked with silver stigma or spots. Grayish hindwings have a darker outer margin, but these are usually hidden when the adult is at rest.

Larva of the Cabbage Looper Moth showing prolegs and anal prolegs.

Tveten noticed that something was eating his swamp sunflowers at his home in Baytown. He went out at night and found 20 mm loopers feeding. These were emerald green with two pairs of prolegs and anal prolegs.

Larvae have two diffuse, whitish dorsolateral stripes with one on each side of the green dorsum. These stripes are "sharp white lateral stripes on each side" per Tveten.

Larvae pupated in a very thin cocoon on the wall of a jar in captivity. Tveten noted that the pupae remained green until the day before emergence. One specimen spent 10 days in the cocoon.

Range in Texas: Most of the state

When Found in Texas: March to December

Food Plants: Swamp sunflower, sage, ornamental salvia, agricultural crops such as cabbage and corn

Soybean Looper Moth   *Chrysodeixis includens*   8890

This dark winged moth has a silver stigma on the forewing. Tveten noted that the spots are separated with whitish spot ahead of silver and forewings have golden highlights. The hindwings are dark.

A moth in captivity laid approximately 200 eggs all over a jar a day after being placed in captivity. The eggs were whitish, or colorless, and have about 32 beaded ribs. The ribs were varying lengths to allow for the shape of the egg.

Larvae emerged four days after eggs were laid. Each was tiny and colorless. By the time the larva was 10 mm long, it was slender and described as "pale, translucent, white (slightly green tinge in some light) with a few hairs on body." The larva had two pairs of prolegs and anal prolegs.

Twelve days after hatching, the larvae were 30 mm long and heavier at the tapered posterior end. Larvae at this stage were lime green with dark spiracles or spots down the sides. A thin yellow stripe ran just below the spiracles with another on the back giving the larva two pairs of stripes.

Cocoon was spun with loose silk net across the top surface of a single leaf. The leaf edges were pulled slightly inward and the caterpillar pupated within. The pupa was pale green and gradually developed brown markings on back of abdomen.

Adult Soybean Looper Moth.

Soybean Looper Moth larva in the lime green stage of development.

Adult emerged nine days after spinning the cocoon.

Range in Texas: Lubbock area to East Texas

When Found in Texas: All year

Food Plant: Dandelion, Peruvian lily, goldenrod, mistflower, soybeans

Adult Pitcher Plant Mining Moth on a pitcher plant.

Pitcher Plant Mining Moth   *Exyra semicrocea*   **9024**

Delta-shaped moth with forewings horizontally divided with white on top and dark gray on lower half. The round hairy head is dark brown. Legs are yellow.

Tveten found this species while exploring a wet ditch in front of the Hazel Green Sanctuary near Hyatt Lake in Tyler County. He noted that it had been a dry year with lots of rain the previous two weeks. Several Grasspink or Calopogon orchids were in bloom along with yellow pitcher plants.

Several caterpillars were found but never more than one per pitcher. Larvae had a looping gait with spines on dorsolateral areas of the middle segment. Most of the larvae were small but four were large and reddish. All spun a light web of silk across the top of the pitcher. Frass dropped down into the tube of the pitcher plant. "Lots of frass," wrote Tveten.

Larvae were feeding on the upper part of the pitcher leaving a membrane remaining.

Tveten found only one caterpillar per pitcher plant.

Tveten performed an experiment with the larvae:

- Two equal-sized larvae were put in a single pitcher plant in three jars. In the first jar, one larva was dead within an hour. In the other two jars, one larva moved to the outside of the pitcher plant. They fed well for three days and then "contracted" unlike the larva inside the pitcher plant.
- On day three, he removed those larvae and gave them their own pitcher plant in a large jar. The larvae remained inside their own pitcher plant.
- Eight days later all the larvae were healthy and webbed in their own pitcher. Most pupated.

Pupae were well down in the tube of the pitcher plants. These were suspended with a few threads; no cocoon. The shed larval skin was present.

Four adult moths were found in the pitcher plants. Two of the adults were mating and the other two were down in a leaf, clinging to the side.

Two years later, Tveten examined pitcher plants at Kirby State Forest in Tyler County. This was in April and one to two eggs were found inside the

Shed larval skin and webbing next to an adult Pitcher Plant Mining Moth.

pitchers. Eggs were the size of a pencil dot with 38–40 ridges around a flattened sphere. Some were "shorter to fat." All were pale yellowish-cream.

Five days after being laid in captivity, eggs turned dark gray. Seven days after being laid the eggs hatched.

Range in Texas: East Texas

When Found in Texas: March to October

Food Plants: Pitcher plants

Lantana Moth   *Diastema tigris*   **9067**

"Kept for my own," wrote Tveten about the Lantana Moth that he found at the Frio River Cabins. He noted that adults are "gorgeous rich brown and gold moths with dark brown, squarish markings." They have a wing-span of about an inch.

The very tiny, round, whitish eggs were laid in a group of 80 to 100. Tveten noted that the eggs hatched in four days.

Larvae were tiny and pale green during the first instar. They fed well on lantana leaves leaving tiny holes but did not feed on the blooms.

Within seven days, the larvae were green with narrow whitish stripes. They rested on the underside of the lantana leaves. When feeding, they skeletonized the leaves leaving only the top membrane.

The last instar was 13 days after hatching. The larvae were green with broken white pattern or chevrons. Tveten noted that they were "loopers, but have all four prolegs." Before pupation, they turned emerald green.

Twenty days after hatching the larvae burrowed in peat moss. Three weeks after pupating, the adults emerged.

Range in Texas: Dallas to Langtry and east in Texas

When Found in Texas: February to November

Food Plants: Lantana

Tveten admired the beauty of the Lantana Moth.

Dagger moths are a large group of unremarkable, gray moths. Their wings are held flat alongside the body in a delta. Wingtips extend past the body and curve to join right above the tail.

Tveten found adult Clear Dagger moths at his porch light in Baytown in March and April. (During his time the moth was called a Prunus Dagger.) He described one as, "pale gray, smeared dark 'dagger' marks." He noted a yellow tuft on the thorax to help identify this species.

One adult was placed in a jar. It laid approximately 150 eggs in two days scattered on the sides and bottom.

Eggs were "very pale green." Tveten's drawing shows a small, flattened bump with minute lateral ridges.

Larvae emerged four days later. All of them emerged at night. Caterpillars were "tiny, hairy, yellowish-white." A larva was found on rattan-vine, also called Alabama supplejack, at Armand Bayou in Harris County. Mature larvae were robust and green with a brown head. Larvae can also be rusty brown.

Larvae fed on cedar elm but refused to eat willow oak or sugar hackberry. The larvae, "ate a small patch on heavier underside of elm leaf,

Adult Clear Dagger moth.

Rusty brown larva of the Clear Dagger moth.

eating just through the epidermis. Leaves are riddled with translucent windows."

At maturity, the larvae were green with a "hunch-backed" brown dorsal patch.

A loose spun cocoon was formed on the bottom of the jar. Adults emerged six weeks after spinning.

Range in Texas: Houston suburbs to Austin and up to Dallas/Fort Worth

When Found in Texas: March to October

Food Plants: Observed cedar elm and rattan vine. Reported apple, cherry, plum, hawthorn.

Adult Eight-Spotted Forester.

Eight-Spotted Forester *Alypia octomaculata* **9314**

Most moths are nocturnal but the eight-spotted forester flies during the day. It is easy to identify with solid black wings. Each wing is adorned with two large white spots, hence its name. The body, if you get to see it, is also black and white. Flashy orange tufts adorn the front legs. Wings are held away from the body like a butterfly.

Tveten found larvae feeding on wild grape and Virginia creeper at the Houston Arboretum in mid-April. In April 1987, he found more larvae feeding on peppervine in Wallisville in East Texas. Returning to Wallisville 11 years later, in April 1998, he found larvae on peppervine once again.

Larvae were 20–26 mm long when found. He described them as "black with white transverse bars and raised orange areas." He noted the "white patch on side rear anal segments" and white hairs or setae. Tveten noted that these were "very beautiful."

Pupation was five to eight days later in dirt covered with sphagnum moss in the bottom of the enclosure.

Larva form of the Eight-Spotted Forester.

Adults emerged 10 to 15 days later. Tveten used "beautiful" and "lovely" to describe the adults. He noted that, "blue iridescent spots on wings (were) visible in some light."

Range in Texas: Panhandle down to Big Bend region and east throughout the state

When Found in Texas: March through November

Food Plants: Wild grape, Virginia creeper, peppervine

EREBINAE—Genus *Catocala* or Underwings

The family Erebidae of moths has a subfamily called the Erebinae. Erebinae subfamily includes the Underwing moths and the Witch moth. Most of these have cryptic coloration to avoid predators.

Underwing moths have a dull forewing that folds over a more colorful hindwing or underwing. Hence the name Underwing moths. The color of the underwing can be bright or dull. It is always a surprise and treat to see.

Adult stage of the Copper Underwing with colorful underwing hidden.

Copper Underwing   *Amphipyra pyramidoides*   **9638**

The underwing of a Copper Underwing is predictably copper colored. The forewings are mottled brown with a lighter band along the trailing edge. The copper hindwings are hidden when the moth is at rest.

Tveten found a Copper Underwing larva at the Big Creek Scenic Area in April 1985. The single larva was on a maple sapling. It measured approximately 35 mm long and was "very heavy." It had a "hump" at the end. Tveten described it as "emerald green with yellow markings. Spiracles— yellow center, raised black in white spot."

The larva formed a cocoon five days after being placed in a container. The cocoon was formed "by cutting a piece from one leaf and stitching it to another."

Five weeks later in May 1985 the adult emerged. The forewings were very dark, and "hindwings are brilliant shiny metallic red-copper." Tveten wrote that the books do not do this species justice.

Range in Texas: Dallas/Fort Worth with scattered reports in the Piney Woods, Austin

When Found in Texas: April to November

Food Plant: Maple, oak, viburnum, grape

Larvae of the Copper Underwing moth have a pronounced "hump" and yellow stripe.

Variations in the color of the adult Yellow-Striped Armyworm Moth.

Yellow-Striped Armyworm Moth *Spodoptera ornithogalli* 93–2219 (9669)

This moth measures less than an inch long. Its mottled brownish wings are held tight to the body. Front legs often extend past the head when clinging to a wall under a porch light.

Tveten found this moth under his porch light in Baytown. While most sightings were in the summer, he did make notes on this moth under the light in December and January. He found it twice in Harlingen under porch lights in November.

Moths laid eggs in a jar a day or two after being placed in captivity. Approximately 100 eggs were laid in a loose grouping on the side of the jar. "Eggs are tiny, round, dark, covered with scales," he wrote. In another instance, the eggs were covered in hairs and secretions. He wrote that the mass "looks like a thick layer of felt. Dark gray mess."

Eggs hatched four to six days later, revealing "extremely tiny greenish larvae." As larvae grew, they were black with yellow and blue lateral stripes. A bottom stripe was red. He noted large black eyespots on sides near the head.

When small, the larvae fed from the underside of a leaf.

Pattern details as larvae of the Yellow-Striped Armyworm Moth as they age.

Twenty days after hatching, the larvae had lost their stripes. They had a dark triangle on the dorsum. Tveten noted that there was a big difference in markings between the fourth and fifth instar.

A larva that measured 24 mm by 5.5 mm was dark brown with lighter dorsum and darker below. The larva was lightly mottled with pale blue dots, particularly on the sides. Behind the head on the first thoracic segment was a pair of dark "eye spots." A different specimen measured 40 mm by 7 mm.

Pupae were on the surface of dirt or slightly buried.

Range in Texas: Entire state with majority of sightings from Dallas, Austin, San Antonio, to the Rio Grande Valley and east

When Found in Texas: All year

Food Plants: St. Augustine grass, rusty-seed paspalum, wild lettuce, violets, oxalis, Chinese privet ligustrum, common sowthistle, swamp sunflower, Maximilian sunflower

Details of the adult Laudable Arches moth.

Laudable Arches *Lacinipolia laudabilis* 93–3065 (10411)

This mottled reddish-brown moth showed up under Tveten's porch light. He called it "very pretty" but was not sure if it was an Implicit Arches, Explicit Arches, or Laudable Arches. Each of these holds their ½-inch forewings along the body to form a rounded delta shape. The laudable arches' forewing is darker in the middle and has a checkered fringe.

Larvae of the Laudable Arches feeding on dandelion.

Larvae of the Laudable Arches feeding on leaves.

Eggs were laid in clusters of 45 to 150. Each was very tiny, white and ridged. The round spheres had a flat bottom, indentation on the top, and radiating ridges.

Eggs hatched six to seven days later. Larvae ate most of the eggshells. They were pale green with large heads and a row of long hairs along their side.

The "armyworm" larvae fed on common dandelion. They were 3–4 mm long 11 days after hatching. A month after hatching, the larvae were 1 inch long and "very uninteresting." Tveten wrote that they were drab with stiff bodies. "Appears dead most of the time," but they "may curl up slightly when disturbed, but not active."

A month after hatching, the larvae were growing well. They fed on dandelion but refused hackberry, willow oak, and Chinese privet.

Range in Texas: Most of Texas except Panhandle and far west

When Found in Texas: All year

Food Plants: Dandelion

Adult of the Adjutant Wainscot moth.

Adjutant Wainscot *Leucania adjuta* 10456

These are small moths with wings held tightly over the body. Their overall shape resembles a pharmacy capsule with cryptic pale straw-color and dark forewings. There is a dark line from the apex of the forewing and a white dot on the forewing.

Tveten called this an "armyworm looper."

Eggs are laid in several masses of hundreds each. Individual eggs are gray and appear smooth.

Larvae hatched seven days after eggs were laid. Each fed on St. Augustine grass. Nearing maturity each was tan and green striped and 1 inch long.

Pupation began 35 days after hatching. Bare pupae rested in the bottom of jars in captivity.

Adult moths emerged 19 days later.

Range in Texas: Eastern half of state from Dallas to Rio Grande Valley, Lubbock area

When Found in Texas: Year round including winter months

Food Plant: St. Augustine grass

Larvae as they feed on grass and mature.

Plump larva of the Adjutant Wainscot before pupating.

Adult Green
Cutworm Moth.

Green Cutworm Moth *Anicla infecta* **10911**

The adult of this species has light tan forewings that are a bit more than
½ inch long. Wings are held back from the body. There is an indistinct
darker spot in the center for the forewing and the submarginal edge is
darker. Adults can also be dark brown with the same markings. Antennae are thin and threadlike.

Larvae were found around the Tveten's home in Baytown.

Eggs are grayish orbs laid on the underside of a leaf. Each egg is spaced
apart from its neighbor.

Young larvae are brown with a lighter stripe on each side. As they
mature, larvae are green with a tan stripe down each side.

When disturbed, they curl into a spiral.

When ready to pupate, the larvae burrowed in dirt. Adults emerged a
month later.

Range in Texas: Statewide

When Found in Texas: All year

Food Plant: Clover, grasses, and tobacco

Variation in the wing colors of the Green Cutworm Moth.

Larva curled into a spiral when disturbed.

Color changes in Green Cutworm Moth larva as they age.

Adult of the Tobacco
Budworm Moth.

Tobacco Budworm Moth *Chloridea virescens* **11071**

This pale, cream colored moth has three lighter diagonal lines on the forewings. Wings are held along the body and form a wide delta. The forewings have a rounded apex.

Tveten found the larva of this species feeding on Turk's cap blooms at the Robert A. Vine Environmental Science Center in Houston. The 37-mm long larva was pale green with broken white lines and a heavier lateral white stripe. There was a reddish wash along each side. Dark red-maroon "humps" each had a hair in the center. The largest "humps" were on abdominal segments one, two, and eight. These were paired on either side of the dorsal line.

Tveten noted that the droppings from the larva were red from the flowers it was eating.

The larva left the plant and wandered "up far." It appeared to be smaller and shriveled. Tveten placed soil with layers of sphagnum in a jar and the larva buried down.

Larva of the Tobacco Budworm feeding on Turk's bloom.

The adult emerged 21 days later.

Range in Texas: Entire state except extreme west

When Found in Texas: April to October

Food Plant: Tobacco pest. Feeds on bud, flowers, and seedpods of nightshade.

Larva of Tobacco Budworm Moth.

Larva to Tobacco Budworm beginning to pupate.

Glossary

apex—pointed or curved tip of the forewing or hindwing

basal—area of the wing closest to the thorax

chrysalis—stage of development or metamorphosis between larva and adult moth

cocoon—casing produced around itself by a moth larva just before changing to a pupa

costa—top leading edge of either the forewing or hindwing of a moth or butterfly

costal edge—the outer edge farthest away from the body on forewing or hindwing of the forewing or hindwing

dimorphic—two distinct forms; for example, male is one color and a female is another, or male is smaller than the female of the same species

discal—the section of the wing next to the basal section; postdiscal follows and the wing edge is the submargin

dorsal—back or upper side

dorsolateral—pertaining to the dorsal and lateral

falcate—curved or hooked in shape

frass—digestive droppings, fecal matter

inner margin—part of the wing usually held back against the thorax

instar—stages in the development of a caterpillar or larva

labial palps—lower lip of an insect; scales or hairs for testing food; or
     scale-covered appendages on the face are part of the mouth and are
     used to sense if something is edible

larva—caterpillar stage of some insects including moths

larvae—plural of larva

larval—one of the life stages of a larva or caterpillar

lateral—pertaining to the side of a moth or caterpillar

median band—center of the wing often demarcated by a color on both
     sides

metamorphosis—insects and amphibians transform from an immature
     form to an adult form and these distinct stages are a metamorphosis;
     metamorphosis in a moth is the change from caterpillar or larva to
     pupa to moth

outer margin—outside edge of the submarginal part of the wing; the
     outer margin of the wing often has a fringe in alternating colors

pectinate antennae—antennae with hair-like segments on one side like a
     comb

postmedial—part of the wing away from the base and closer to the outer
     margin; usually a location for a line or darker/lighter area on the wing

pronotum—plate-like structure on the thorax of an insect

pupa—stage of development or metamorphosis between larva and adult
     moth or butterfly; pupa is naked and hard

pupae—plural of pupa

reniform spot—kidney-shaped spot or shape on the forewing of a moth;
     usually in the center

spiracles—openings on the side of a larva's body; part of the respiration system but usually this area is colored and is one of our field marks

submarginal—part of the wing before the outer margin

tubercles—protuberance that is usually small and rounded

ventral—underside

Index

Note: Illustrations are indicated by *italic* page numbers.

stealth-fighter shape: of faint-spotted palthis, 221, 222; of Tersa sphinx, 150, 151

stinging larvae (asps), 29, 30, 32–33, 96, 97

submarginal, definition of, 275

sunflowers: cabbage looper and, 245; moonseed moth and, 3

sweetgum: fall webworm and, 195; luna moth and, 112–14

Synchlora frondaria. See southern emerald

Syssphinx heiligbrodti. See Heiligbrodt's mesquite moth

Syssphinx hubbardi. See Hubbard's small silkmoth

taxonomy of moths, 13, 20

tent caterpillars: eastern, 68–70, 68–70; forest, 65–67, 65–67; western, 71, 71–72

Tersa sphinx, 150–56; adult, 132, 150, 151, 157; as hummingbird mimic, 150–56; larvae of, 150, 152–53, 156; pupae of, 150, 154–55, 158; Tveten's description of, 150

Texas A&M University, Tveten's papers and photographs at, 11, 19

Texas mocis, 234

Texas Parks & Wildlife (magazine): posthumous tribute to Tveten in, 6–11; Tveten's work in, 9

threadlike antennae, 13, 14

Thyridopteryx ephemeraeformis. See evergreen bagworm

tiger moths: fall webworm *versus*, 197; Nevada, 206–7, 206–7; Virginian, 191–93, 191–94

Titan sphinx, Tersa sphinx mistaken for, 150–56

tobacco budworm moth, 270–71; adult, 270, 270; larvae of, 270, 271–72

tobacco hornworm, 134; adult, 134, 134; larvae of, 134, 135

tomato plants, tobacco hornworm and, 134

tree-of-heaven, 26

Trichoplusia ni. See cabbage looper moth

tubercles: of Calleta silkmoth, 117; of curve-tooth geometer, 49; definition of, 275; of Io moth, 104–7; of luna moth, 113; of Nanina oak-slug moth, 39; of Promethea moth, 122

tussock moths: classic "tufts" of, 216–17; Davis', 20, 208, 208; Kerrville, 217, 219–20; milkweed, 212, 212–13; Schaus', 20, 209–11, 209–11; southern, 216–17, 216–18

Tveten, Gloria, 5; books co-authored by, 8, 9; joint byline for newspaper column, 9; participation in fieldwork and research, 3–4, 7; relationship with author, 5

Tveten, John, 3–11, 5; book based on research of, 3, 11, 16; boyhood of, 8; career change of, 3, 8–9; collection and observation methods of, 4–5, 16–18; as consummate naturalist, 5–11; education of, 8; experience of walking with, 6–7; field notes of, 3–4, 4, 11, 20; influence of Houston lecture, 7–8; newspaper column of, 9; photography of, 7,